T0269283

SpringerBriefs in Intelligent Systems

Artificial Intelligence, Multiagent Systems, and Cognitive Robotics

This series covers the entire research and application spectrum of intelligent systems, including artificial intelligence, multiagent systems, and cognitive robotics. Typical texts for publication in the series include, but are not limited to, state-of-the-art reviews, tutorials, summaries, introductions, surveys, and in-depth case and application studies of established or emerging fields and topics in the realm of computational intelligent systems. Essays exploring philosophical and societal issues raised by intelligent systems are also very welcome.

More information about this series at http://www.springer.com/series/11845

Frans A. Oliehoek · Christopher Amato

A Concise Introduction
to Decentralized POMDPs

 Springer

Frans A. Oliehoek
School of Electrical Engineering, Electronics
    and Computer Science
University of Liverpool
Liverpool
UK

Christopher Amato
Computer Science and Artificial Intelligence
    Laboratory
MIT
Cambridge, MA
USA

ISSN 2196-548X   ISSN 2196-5498   (electronic)
SpringerBriefs in Intelligent Systems
ISBN 978-3-319-28927-4   ISBN 978-3-319-28929-8   (eBook)
DOI 10.1007/978-3-319-28929-8

Library of Congress Control Number: 2016941071

Printed on acid-free paper

This Springer imprint is published by Springer Nature
The registered company is Springer International Publishing AG Switzerland

Dedicated to future generations of intelligent decision makers.

Preface

This book presents an overview of formal decision making methods for decentralized cooperative systems. It is aimed at graduate students and researchers in the fields of artificial intelligence and related fields that deal with decision making, such as operations research and control theory. While we have tried to make the book relatively self-contained, we do assume some amount of background knowledge.

In particular, we assume that the reader is familiar with the concept of an *agent* as well as search techniques (like depth-first search, A*, etc.), both of which are standard in the field of artificial intelligence [Russell and Norvig, 2009]. Additionally, we assume that the reader has a basic background in probability theory. Although we give a very concise background in relevant single-agent models (i.e., the 'MDP' and 'POMDP' frameworks), a more thorough understanding of those frameworks would benefit the reader. A good first introduction to these concepts can be found in the textbook by Russell and Norvig, with additional details in texts by Sutton and Barto [1998], Kaelbling et al. [1998], Spaan [2012] and Kochenderfer et al. [2015]. We also assume that the reader has a basic background in game theory and game-theoretic notations like Nash equilibrium and Pareto efficiency. Even though these concepts are not central to our exposition, we do place the Dec-POMDP model in the more general context they offer. For an explanation of these concepts, the reader could refer to any introduction on game theory, such as those by Binmore [1992], Osborne and Rubinstein [1994] and Leyton-Brown and Shoham [2008].

This book heavily builds upon earlier texts by the authors. In particular, many parts were based on the authors' previous theses, book chapters and survey articles [Oliehoek, 2010, 2012, Amato, 2010, 2015, Amato et al., 2013]. This also means that, even though we have tried to give a relatively complete overview of the work in the field, the text in some cases is biased towards examples and methods that have been considered by the authors. For the description of further topics in Chapter 8, we have selected those that we consider important and promising for future work. Clearly, there is a necessarily large overlap between these topics and the authors' recent work in the field.

Acknowledgments

Writing a book is not a standalone activity; it builds upon all the insights developed in the interactions with peers, reviewers and coauthors. As such, we are grateful for the interaction we have had with the entire research field. We specifically want to thank the attendees and organizers of the workshops on *multiagent sequential decision making (MSDM)* which have provided a unique platform for exchange of thoughts on decision making under uncertainty.

Furthermore, we would like to thank João Messias, Matthijs Spaan, Shimon Whiteson, and Stefan Witwicki for their feedback on sections of the manuscript.

Finally, we are grateful to our former supervisors, in particular Nikos Vlassis and Shlomo Zilberstein, who enabled and stimulated us to go down the path of research on decentralized decision making.

Contents

Acronyms

AH   action history
AOH   action-observation history
BG   Bayesian game
CG   coordination graph
CBG   collaborative Bayesian game
COP   constraint optimization problem
DAG   directed acyclic graph
DP   dynamic programming
Dec-MDP   decentralized Markov decision process
Dec-POMDP   decentralized partially observable Markov decision process
DICE   direct cross-entropy optimization
EM   expectation maximization
EXP   deterministic exponential time (complexity class)
FSC   finite-state controller
FSPC   forward-sweep policy computation
GMAA*   generalized multiagent A*
I-POMDP   interactive partially observable Markov decision process
MAS   multiagent system
MARL   multiagent reinforcement learning
MBDP   memory-bounded dynamic programming
MDP   Markov decision process
MILP   mixed integer linear program
NEXP   non-deterministic exponential time (complexity class)
ND-POMDP   networked distributed POMDP
NDP   nonserial dynamic programming
NLP   nonlinear programming
NP   non-deterministic polynomial time (complexity class)
OH   observation history
POMDP   partially observable Markov decision process
PSPACE   polynonomial SPACE (complexity class)
PWLC   piecewise linear and convex

RL   reinforcement learning
TD-POMDP   transition-decoupled POMDP

List of Symbols

Throughout this text, we tried to make consistent use of typesetting to convey the meaning of used symbols and formulas. In particular, we use blackboard bold fonts ($\mathbb{A}$,$\mathbb{B}$, etc.) to denote sets, and subscripts to denote agents (typically i or j) or groups of agents, as well as time (t or τ).

For instance a is the letter used to indicate actions in general, a_i denotes an action of agent i, and the set of its actions is denoted $\mathbb{A}_i$. The action agent i takes at a particular time step t is denoted $a_{i,t}$. The profile of actions taken by all agents, a joint action, is denoted a, and the set of such joint actions is denoted $\mathbb{A}$. When referring to the action profile of a subset e of agents we write a_e, and for the actions of all agents except agent i, we write a_{-i}. On some occasions we will need to indicate the index within a set, for instance the k-th action of agent i is written a_i^k. In the list of symbols below, we have shown all possible uses of notation related to actions (base symbol 'a'), but have not exhaustively applied such modifiers to all symbols.

$\cdot$	multiplication,
$\times$	Cartesian product,
$\circ$	policy concatenation,
$\Downarrow$	subtree policy consumption operator,
$\triangle(\cdot)$	simplex over $(\cdot)$,
$\mathbf{1}_{\{\cdot\}}$	indicator function,
β	macro-action termination condition,
Γ_j	mapping from histories to subtree policies,
$\Gamma^{\mathscr{X}}$	state factor scope backup operator,
$\Gamma^{\mathscr{A}}$	agent scope backup operator,
γ	discount factor,
δ_t	decision rule for stage t,
δ_t	joint decision rule for stage t,
$\hat{\delta}_t$	approximate joint decision rule,
$\delta_{i,t}$	decision rule for agent i for stage t,
Δt	length of a stage ts,

ε	(small) constant,
$\bar{\theta}$	joint action-observation history,
$\bar{\theta}_i$	action-observation history,
$\bar{\Theta}_i$	action-observation history set,
l_i	information state, or belief update, function,
μ_i	macro-action policy for agent i,
π	joint policy,
π_i	policy for agent i,
π_{-i}	(joint) policy for all agents but i,
π^*	optimal joint policy,
ρ	number of reward functions,
Σ	alphabet of communication messages,
σ_t	plan-time sufficient statistic,
τ	stages-to-go ($\tau = h - 1$),
υ	domination gap,
Φ_{Next}	set of next policies,
φ_t	past joint policy,
ξ	parameter vector,
ψ	correlation device transition function,
$\mathbb{A}$	set of joint actions,
$\mathbb{A}_i$	set of actions for agent i,
$\bar{\mathbb{A}}$	joint action history set,
$\bar{\mathbb{A}}_i$	action history set for agent i,
a	joint action,
a_t	joint action at stage t,
a_i	action for agent i,
a_e	joint action for subset e of agents,
a_{-i}	joint action for all agents except i,
$\bar{a}_i$	action history of agent i,
$\bar{a}_{i,t}$	action history of agent i at stage t,
$\bar{a}$	joint action history,
$\bar{a}_t$	joint action history at stage t,
$B(b_0, \varphi_t)$	Bayesian game for a stage,
$B(\mathcal{M}_{DecP}, b_0, \varphi_t)$	CBG for stage t of a Dec-POMDP,
$\mathbb{B}$	set of joint beliefs,
b_0	initial state distribution,
$b_i(s_t, q_{-i}^\tau)$	multiagent belief,
b	belief,
b_i	belief for agent i (e.g., a multiagent belief),
$\mathbb{C}$	states of a correlation device,
C_Σ	message cost function,
c	correlation device state,
$\mathbb{D}$	set of agents,
$\mathbf{E}[\cdot]$	expectation of $\cdot$,

$\mathscr{E}$	set of hyper-edges,
e	hyper edge, or index of local payoff function (corresponding to a hyper edge),
f_ξ	probability distribution, parameterized by ξ,
$f_{\xi(j)}$	distribution over joint policies at iteration j,
h	horizon,
$I_{i \to j}$	influence of agent i on agent j,
I_i	information states for agent i,
$\mathbb{I}_i$	set of information states for agent i,
$\mathcal{M}$	Markov multiagent environment,
$\mathcal{M}_{DecP}$	Dec-POMDP,
$\mathcal{M}_{MPOMDP}$	MPOMDP,
$\mathcal{M}_{PT}$	plan-time NOMDP,
m_i	agent model, also finite-state controller,
m	agent component (a joint model),
m_i	macro-action for agent i,
N_b	number of best samples,
N_f	number of fire levels,
Next	operation constructing next set of partial policies,
$NULL$	null observation,
n	number of agents,
O	observation function,
O_i	local observation function for agent i,
OC	optimality criterion,
$\mathbb{O}$	set of joint observations,
$\mathbb{O}_i$	set of observations for agent i,
$\bar{\mathbb{O}}$	joint observation history set,
$\bar{\mathbb{O}}_i$	observation history set for agent i,
o	joint observation,
o_i	observation for agent i,
$o_{i,\emptyset}$	NULL observation for agent i,
$\bar{o}_i$	observation history of agent i,
$\bar{o}_t$	joint observation history at stage t,
$\bar{o}_{t,|k|}$	joint observation history at stage t of length k,
Q^π	Q-value function for π,
$\mathbb{Q}_i^\tau$	set of q_i^τ,
$\mathbb{Q}^\tau$	set of q^τ,
$\mathbb{Q}_{e,i}^{\tau+1}$	set of subtree policies resulting from exhaustive backup,
$\mathbb{Q}_{m,i}^{\tau+1}$	set of maintained subtree policies,
q_{t-k}^k	joint subtree policy of length k to be executed at stage $t-k$,
q_i^τ	τ-stage-to-go subtree policy for agent i,
q^τ	τ-stage-to-go joint subtree policy,
R	reward function,
R_i	local reward function for agent i,
R^e	local reward function (with index e),
$\mathbb{R}$	real numbers,

$\mathbb{S}$ state space (the set of all states),

$\mathbb{S}_i$ set of local state for agent i,

$\check{\mathbb{S}}_i$ set of interactive states for agent i,

s state,

s_e local state of agents participating in e (in agent-wise factored model),

s_i local state for agent i,

T transition function,

T_i local state transition function for agent i,

t stage,

$U_{ss}()$ sufficient statistic update,

u payoff function (in context of single-shot game),

u^e local payoff function,

V value function,

V^* optimal value function,

V^e value function for a particular payoff component e,

V^π value function for joint policy π,

v value vector,

v_δ value vector associated with (meta-level 'action') δ,

$\mathcal{V}$ set of value vectors,

$\mathcal{X}$ space of candidate solutions (DICE),

$\mathbf{X}$ set of samples (DICE),

$\mathbf{X}_b$ set of best samples (DICE),

$\mathbb{X}$ set of state factors

$\mathbb{X}^j$ set of values for j-th state factor,

x^j value for j-th state factor,

x^H fire level of house H,

x_e profile of values for state factors in component e,

x candidate solution (DICE),

$\mathbb{Z}_i$ set of auxiliary observations for agent i,

Chapter 1
Multiagent Systems Under Uncertainty

The impact of the advent of the computer on modern societies can hardly be over-stated; every single day we are surrounded by more devices equipped with on-board computation capabilities taking care of the ever-expanding range of functions they perform for us. Moreover, the decreasing cost and increasing sophistication of hardware and software opens up the possibility of deploying a large number of devices or systems to solve real-world problems. Each one of these systems (e.g., computer, router, robot, person) can be thought of as an *agent* which receives information and makes decisions about how to act in the world. As the number and sophistication of these agents increase, controlling them in such a way that they consider and cooperate with each other becomes critical. In many of these *multiagent systems (MASs)*, cooperation is made more difficult by the fact that the environment is unpredictable and the information available about the world and other agents (through sensors and communication channels) is noisy and imperfect. Developing agent controllers by hand becomes very difficult in these complex domains, so automated methods for generating solutions from a domain specification are needed. In this book, we describe a formal framework, called the *decentralized partially observable Markov decision process (Dec-POMDP)*, that can be used for decision making for a team of cooperative agents. Solutions to Dec-POMDPs optimize the behavior of the agents while considering the uncertainty related to the environment and other agents. As discussed below, the Dec-POMDP model is very general and applies to a wide range of applications.

From a historical perspective, thinking about interaction has been part of many different 'fields' or, simply, aspects of life, such as philosophy, politics and war. Mathematical analyses of problems of interaction date back to at least the beginning of the eighteenth century [Bellhouse, 2007], driven by interest in games such as chess [Zermelo, 1913]. This culminated in the formalization of game theory with huge implications for the field of economics since the 1940s [von Neumann and Morgenstern, 1944]. Other single-step cooperative team models were studied by Marschak [1955] and Radner [1962], followed by systems with dynamics modeled as team theory problems [Marschak and Radner, 1972, Ho, 1980] and the resulting complexity analysis [Papadimitriou and Tsitsiklis, 1987]. In the 1980s, people

© The Author(s) 2016
F.A. Oliehoek and C. Amato, *A Concise Introduction to Decentralized POMDPs*,
SpringerBriefs in Intelligent Systems, DOI 10.1007/978-3-319-28929-8_1

in the field of Artificial Intelligence took their concept of *agent*, an artificial entity that interacts with its environment over a sequence of time steps, and started thinking about multiple such agents and how they could interact [Davis and Smith, 1983, Grosz and Sidner, 1986, Durfee et al., 1987, Wooldridge and Jennings, 1995, Tambe, 1997, Jennings, 1999, Lesser, 1999]. Dec-POMDPs represent a probabilistic generalization of this multiagent framework to model uncertainty with respect to outcomes, environmental information and communication. We first discuss some motivating examples for the Dec-POMDP model and then provide additional details about multiagent systems, the uncertainty considered in Dec-POMDPs and application domains.

1.1 Motivating Examples

Before diving deeper, this section will present two motivating examples for the models and techniques described in this book. The examples briefly illustrate the difficulties and uncertainties one has to deal with when automating decisions in real-world decentralized systems. Several other examples and applications are discussed in Section 1.4.

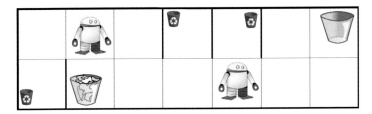

Fig. 1.1: Illustration of a simple Recycling Robots example, in which two robots remove trash in an office environment with three small (blue) trash cans and two large (yellow) ones. In this situation, the left robot may observe that the large trash can next to it is full, and the other robot may detect that the adjacent small trash can is empty. Note that neither robot can be sure of the trash can's true state due to limited sensing capabilities, nor do the robots see the state of trash cans further away. Also, the robots cannot observe each other at this distance and they do not know the observations of the other robot due to a lack of communication.

Multirobot Coordination: Recycling Robots   Consider a team of robots tasked with removing trash from an office building, depicted in Figure 1.1. The robots have sensors to find marked trash cans, motors to move around in order to look for cans, as well as gripper arms to grasp and carry a can. Small trash cans are light and compact enough for a single robot to carry, but large trash cans require multiple robots to carry them out together. It is not certain exactly where the trash cans may be or how fast they will fill up, but it is known that, because more people use them, the larger trash cans fill up more quickly. Each robot needs to take actions based on its own knowledge: while we encourage the robots to share some important information, we would not want them to communicate constantly, as this could overload the office's wireless network and seems wasteful from an energy perspective. Each robot must also ensure that its battery remains charged by moving to a charging station before it expires. The battery level for a robot degrades due to the distance the robot travels and the weight of the item being carried. Each robot only knows its own battery level (but not that of the other robots) and the location of other robots within sensor range. The goal of this problem is to remove as much trash as possible in a given time period. To accomplish this goal we want to find a plan, or policy, that specifies for each robot how to behave as a function of its own observations, such that the joint behavior is optimal. While this problem may appear simple, it is not. Due to uncertainty, the robots cannot accurately predict the amount of battery reduction that results from moving. Furthermore, due to noisy and insufficient sensors, each robot does not accurately know the position and state of the trash cans and other robots. As a result of these information deficiencies, deciding which trash cans to navigate to and when to recharge the battery is difficult. Moreover, even if hand-coding the solution for a single robot would be feasible, predicting how the combination of policies (one for each robot) would perform in practice is extremely challenging.

Efficient Sensor Networks

Another application that has received significant interest over the last two decades is that of sensor networks. These are networks of sensors (the agents) that are distributed in a particular environment with the task of measuring certain things about that environment and distilling this into high-level information. For instance, one could think about sensor networks used for air pollution monitoring [Khedo et al., 2010], gas leak detection [Pavlin et al., 2010], tracking people in office environments [Zajdel et al., 2006, Satsangi et al., 2015] or tracking of wildlife [Garcia-Sanchez et al., 2010]. Successful application of sensor networks involves answering many questions, such as what hardware to use, how the information from different sensors can be fused, and how the sensors should measure various parts of their environment to maximize information while minimizing power use. It is especially questions of the latter type, which involve local decisions by the different sensors, for which the Dec-POMDP framework studied in this book is relevant: in order to decide about when to sense at a specific sensor node we need to reason about the expected information gain from turning that sensor on, which depends on the actions taken at other sensors, as well as how the phenomenon to be tracked moves through the spatial environment. For example, when tracking a person in an office environment, it may be sufficient to only turn on a sensor at the location where the target is expected given all the previous observations in the entire system. Only when the target is not where it is expected to be might other sensors be needed. However, when communication bandwidth or energy concerns preclude the sharing of such previous information, deciding when to turn on or not is even further complicated. Again, finding plans for such problems is highly nontrivial: even if we were able to specify plans for each node by hand, we typically would not know how good the joint behavior of the sensor network is, and whether it could be improved.

Concluding, we have seen that in both these examples there are many different aspects such as decentralization and uncertainties that make it very difficult to specify good actions to take. We will further elaborate on these issues in the remainder of this introductory chapter and give an overview of many more domains for which Decentralized POMDPs are important in Section 1.4.

1.2 Multiagent Systems

This book focuses on settings where there are multiple decision makers, or *agents*, that jointly influence their environment. Such an environment together with the multiple agents that operate in it is called a *multiagent system (MAS)*. The field of MAS research is a broad interdisciplinary field with relations to distributed and concurrent systems, artificial intelligence (AI), economics, logic, philosophy, ecology and the social sciences [Wooldridge, 2002]. The subfield of AI that deals with principles and design of MASs is also referred to as 'distributed AI'. Research on MASs is motivated by the fact that it can potentially provide [Vlassis, 2007, Sycara, 1998]:

- Speedup and efficiency, due to the asynchronous and parallel computation.
- Robustness and reliability, since the whole system can undergo a 'graceful degradation' when one or more agents fail.
- Scalability and flexibility, by adding additional agents as required.
- Lower cost, assuming the agents cost much less than a centralized system.
- Lower development cost and reusability, since it is easier to develop and maintain a modular system.

There are many different aspects of multiagent systems, depending on the type of agents, their capabilities and their environment. For instance, in a homogeneous MAS all agents are identical, while in a heterogeneous MAS the design and capabilities of each agent can be different. Agents can be cooperative, self-interested or adversarial. The environment can be dynamic or static. These are just a few of many possible parameters, leading to a number of possible settings too large to describe here. For a more extensive overview, we refer the reader to the texts by Huhns [1987], Singh [1994], Sycara [1998], Weiss [1999], Stone and Veloso [2000], Yokoo [2001], Wooldridge [2002], Bordini et al. [2005], Shoham and Leyton-Brown [2007], Vlassis [2007], Buşoniu et al. [2008] and Weiss [2013]. In this book, we will focus on decision making for heterogeneous, fully cooperative MASs in dynamic, uncertain environments in which agents need to act based on their individual knowledge about the environment. Due to the complexity of such settings, hand-coded solutions are typically infeasible for such settings [Kinny and Georgeff, 1997, Weiss, 2013]. Instead, the approach advocated in this book is to describe such problems using a formal model—the *decentralized partially observable Markov decision process (Dec-POMDP)*—and to develop automatic decision making procedures, or *planning methods*, for them.

Related Approaches We point out that due to the multi-disciplinary nature of the field of MASs, there are many disciplines that are closely related to the topic at hand, and we point out the most relevant of them here.

For instance, in the field of game theory, much research focuses on extensive form games and partially observable stochastic games, both of which are closely related to Dec-POMDPs (more on this connection in Section 2.4.5). The main difference is that game theorists have typically focused on self-interested settings.

The 'classical planning problem' as studied in the AI community also deals with decision making, but for a single agent. These methods have been extended to the multiagent setting, resulting in a combination of planning and coordination, e.g. distributed problem solving (DPS) [Durfee, 2001]. However, like classical planning itself, these extensions typically fail to address stochastic or partially observable environments [desJardins et al., 1999, de Weerdt et al., 2005, de Weerdt and Clement, 2009].

The field of teamwork theory also considers cooperative MAS, and the *belief-desire-intension (BDI)* model of practical reasoning [Bratman, 1987, Rao and Georgeff, 1995, Georgeff et al., 1999] has inspired many teamwork theories, such as *joint intentions* [Cohen and Levesque, 1990, 1991a,b] and *shared plans* [Grosz and Sidner, 1990, Grosz and Kraus, 1996], and implementations [Jennings, 1995, Tambe, 1997, Stone and Veloso, 1999, Pynadath and Tambe, 2003]. While such BDI-based approaches do allow for uncertainty, they typically rely on (manually) pre-specified plans that might be difficult to specify and have as a drawback the fact that it is difficult to define clear quantitative measures for their performance, making it hard to judge their quality [Pynadath and Tambe, 2002, Nair and Tambe, 2005].

Finally, there are also close links to the operations research (OR) and control theory communities. The Dec-POMDP model is a generalization of the (single-agent) MDP [Bellman, 1957] and POMDP [Åström, 1965] models which were developed in OR, and later became popular in AI as a framework for planning for agents [Kaelbling et al., 1996, 1998]. Control theory and especially optimal control essentially deals with the same type of planning problems, but with an emphasis on continuous state and action spaces. Currently, researchers in the field of decentralized control are working on problems very similar to Dec-POMDPs [e.g., Nayyar et al., 2011, 2014, Mahajan and Mannan, 2014], and, in fact, some results have been established in parallel both in this and the AI community.

1.3 Uncertainty

Many real-world applications for which we would want to use MASs are subject to various forms of uncertainty. This makes it difficult to predict the outcome of a particular plan (e.g., there may be many possible outcomes) and thus complicates finding good plans. Here we discuss different types of uncertainty that the Dec-POMDP framework can cope with.

Outcome Uncertainty.   In many situations, the outcome or effects of actions may be uncertain. In particular we will assume that the possible outcomes of an action are known, but that each of those outcomes is realized with some probability (i.e., the state of the environment changes stochastically). For instance, due to different surfaces leading to varying amounts of wheel slip, it may be difficult, or even impossible, to accurately predict exactly how far our recycling robots move. Similarly, the amount of trash being put in a bin depends on the activities performed by the humans in the environment and is inherently stochastic from the perspective of any reasonable model.[1]

State Uncertainty.   In the real world an agent might not be able to determine what the state of the environment exactly is. In such cases, we say that the environment is *partially observable*. Partial observability results from noisy and/or limited sensors. Because of *sensor noise* an agent can receive faulty or inaccurate sensor readings, or *observations*. For instance, the air pollution measurement instruments in a sensor network may give imperfect readings, or gas detection sensors may fail to detect gas with some probability. When sensors are *limited*, the agent is unable to observe the differences between certain states of the environment because they inherently cannot be distinguished by the sensor. For instance, a recycling robot may simply not be able to tell whether a trash can is full if it does not first navigate to it. Similarly, a sensor node typically will only make a local measurement. Due to such sensor limitations, the same sensor reading might require different action choices, a phenomenon referred to as *perceptual aliasing*. In order to mitigate these problems, an agent may use the history of actions it took and the observations it made to get a better estimate of the state of the environment.

Multiagent Uncertainty: Uncertainty with Respect to Other Agents.   Another complicating factor in MASs is the presence of multiple agents that each make decisions that influence the environment. The difficulty is that each agent can be uncertain regarding the other agents' actions. This is apparent in self-interested and especially adversarial settings, such as games, where agents may not share information or try to mislead other agents [Binmore, 1992]. In such settings each agent should try to accurately predict the behavior of the others in order to maximize its payoff. But even in cooperative settings, where the agents have the same goal and therefore are willing to coordinate, it is nontrivial how such coordination should be

---

[1] To be clear, here we exclude models that try to predict human activities in a deterministic fashion (e.g., this would require perfectly modeling the current activities as well as the 'internal state' of all humans in the office building) from the set of reasonable models.

performed [Boutilier, 1996]. Especially when communication capabilities are limited or absent, the question of how the agents should coordinate their actions is problematic (e.g., given the location of the other recycling robot, should the first robot move to the trash can which is known to be full?). This problem is magnified in partially observable environments: as the agents are not assumed to observe the complete state of the environment—each agent only knows its own observations made and actions taken—there is no common signal that they can use to condition their actions on (e.g., given that the first robot only knows that a trash can was not full four hours ago, should it check if it is full now?). Note that this problem is in addition to the problem of partial observability, and not a substitute for it; even if the agents could freely and instantaneously communicate their individual observations, the joint observations would in general not disambiguate the true state of the environment.

Other Forms of Uncertainty. We note that in this book we build upon the framework of probability to represent the aforementioned uncertainties. However, there are other manners by which one can represent uncertainty, such as Dempster-Shafer belief functions, possibility measures, ranking functions and plausibility measures [Halpern, 2003]. In particular, many of these alternatives try to overcome some of the shortcomings of probability in representing uncertainty. For instance, while probability can represent that the outcome of a die roll is uncertain, it requires us to assign numbers (e.g., $1/6$) to the potential outcomes. As such, it can be difficult to deal with settings where these numbers are simply not known. For a further discussion on such issues and alternatives to probability, we refer you to Halpern [2003].

1.4 Applications

Decision making techniques for cooperative MASs under uncertainty have a great number of potential applications, ranging from more abstract tasks located in a digital or virtual environment to a real-world robotics setting. Here we give an overview of some of these.

An example of a more abstract task is **distributed load balancing** among queues. Here, each agent represents a processing unit with a queue, and can only observe its own queue size and that of its immediate neighbors. The agents have to decide whether to accept new jobs or pass them to another queue. Such a restricted problem can be found in many settings, for instance, industrial plants or a cluster of webservers. The crucial difficulty is that in many of these settings, the overhead associated with communication is too high, and the processing units will need to make decisions on local information [Cogill et al., 2004, Ouyang and Teneketzis, 2014].

Another abstract, but very important domain is that of transmission protocols and routing in **communication networks**. In these networks, the agents (e.g., routers) operate under severe communication restrictions, since the cost of sending meta-level communication (occupying the communication channels) is prohibitively large. For example, consider the problem faced by transmission protocols such as TCP: when should packets be sent across a shared channel to be both fair and efficient for the endpoint computers? Because each computer only has information such as the number of packets in its queue or the latency of its own packets, each computer must make decisions based on its own information. A simple two-agent networking example was first modeled as a Dec-POMDP by Bernstein et al. [2005], but more recently, more realistic congestion control problems have been studied in simulation [Winstein and Balakrishnan, 2013]. Peshkin [2001] treated a packet routing application in which agents are routers and have to minimize the average transfer time of packets. They are connected to immediate neighbors and have to decide at each time step to which neighbor to send each packet. Some other approaches to modeling and optimizing communication networks using decentralized, stochastic, partially observable systems are given by Ooi and Wornell [1996], Tao et al. [2001] and Altman [2002].

The application domain of **sensor networks** [Lesser et al., 2003], as mentioned above, had received much attention in the Dec-POMDP community. These problems are inherently partial observable and—when assuming that adding extensive communication infrastructure between the sensor nodes is infeasible—decentralized. In addition the systems they are intended to monitor are seldom deterministic. This means that these domains have all the facets of a Dec-POMDP problem. However, there are also some special properties, such as the fact that the nodes usually have a static location and that they typically[2] do not change the environment by monitoring, that make them easier to deal with, as we will discuss in Chapter 8.

Another interesting potential application area is the **control of networks of traffic lights**. Smart traffic light controllers have the potential to significantly increase the throughput of traffic [Van Katwijk, 2008], but controlling networks of traffic lights is still challenging, since the traffic flows are stochastic and the networks are large. To avoid a central point of failure and expensive communication infrastructure, the traffic lights should make decisions based on local information, but reasoning about non-local effects and interactions is necessary. A number of papers address such problems from the learning perspective [Wiering, 2000, Wiering et al., 2004, Bazzan, 2005, Kuyer et al., 2008, Bazzan et al., 2010]. Wu et al. [2013] presented a simplified Dec-POMDP traffic control benchmark. The structure present in such traffic problems is similar to the structure exploited by several recent solution methods [Oliehoek et al., 2015b].

A very exciting application domain is that of **cooperative robotics** [Arai et al., 2002]. Robotic systems are notorious for being complicated by stochasticity, sensor noise and perceptual aliasing, and not surprisingly many researchers have used POMDPs to address these problems [Roy et al., 2003, Pineau and Gordon, 2005, Theocharous et al., 2004, Smith, 2007, Kaelbling and Lozano-Pérez, 2013]. In case of multiple cooperative robots, as in the RECYCLING ROBOTS example, the POMDP model no longer suffices; in most of these domains full communication is

---

[2] This assumes the absence of the so-called *observer effect,* as present in quantum mechanics.

either not possible (e.g., there is too little bandwidth to transmit video streams from many cameras or transmission is not sufficiently powerful) or consumes resources (e.g., battery power) and thus has a particular cost. Therefore Dec-POMDPs are crucial for essentially all teams of embodied agents. Examples of such settings are considered both in theory/simulation, such as multirobot space exploration [Becker et al., 2004b, Witwicki and Durfee, 2010b], as well as in real hardware robot implementation, e.g., multirobot search of a target [Emery-Montemerlo, 2005], robotic soccer [Messias, 2014] and a physical implementation of a problem similar to RECYCLING ROBOTS [Amato et al., 2014].

A final, closely related, application area is that of **decision support systems** for complex real-world settings, such as crisis management. Also in this setting, it is inherently necessary to deal with the real world, which often is highly uncertain. For instance, a number of research efforts have been performed within the context of RoboCup Rescue [Kitano et al., 1999]. In particular, researchers have been able to model small subproblems using Dec-POMDPs [Nair et al., 2002, 2003a,b, Oliehoek and Visser, 2006, Paquet et al., 2005]. Another interesting application is presented by Shieh et al. [2014], who apply Dec-MDPs in the context of security games which have been used for securing ports, airports and metro-rail systems [Tambe, 2011].

Chapter 2
The Decentralized POMDP Framework

In this chapter we formally define the Dec-POMDP model. It is a member of the family of discrete-time planning frameworks that are derived from the single-agent Markov decision process. Such models specify one or more agents that inhabit a particular environment, which is considered at discrete *time steps*, also referred to as *stages* [Boutilier et al., 1999] or *(decision) epochs* [Puterman, 1994]. The number of time steps during which the agents will interact with their environment is called the *horizon* of the decision problem, and will be denoted by h.

The family of MDP-derived frameworks considered in decision-theoretic planning very neatly fits the definition of an agent [Russell and Norvig, 2009] by offering an interface of actions and observations to interact with the environment. At each stage $t = 0, 1, 2, \ldots, h - 1$ every agent under consideration takes an action and the combination of these actions influences the environment, causing a state transition. At the next time step, each agent first receives an observation of the environment, after which it has to take an action again. The way in which the environment changes and emits observations is modeled by the transition and observation model. These specify probabilities that represent the stochastic dynamics of the environment. Additionally there are rewards that specify what behavior is desirable. Hence, the reward model defines the agents' goal or task: the agents have to come up with a plan that maximizes the expected long-term reward signal.

2.1 Single-Agent Decision Frameworks

Before diving into the core of multiagent decision making under uncertainty, we first give a concise treatment of the single-agent problems that we will build upon. In particular, we will treat *Markov decision processes (MDPs)* and *partially observable Markov processes (POMDPs)*. We expect the reader to be (somewhat) familiar with these models. Hence, these sections serve as a refresher and to introduce notation. For more details we refer the reader to the texts by Russell and Norvig [2009], Puterman [1994], Sutton and Barto [1998], Kaelbling et al. [1998] and Spaan [2012].

© The Author(s) 2016
F.A. Oliehoek and C. Amato, *A Concise Introduction to Decentralized POMDPs*,
SpringerBriefs in Intelligent Systems, DOI 10.1007/978-3-319-28929-8_2

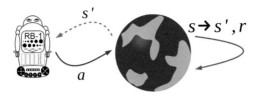

Fig. 2.1: Schematic representation of an MDP. At every stage, the agent takes an action and observes the resulting state s'.

2.1.1 MDPs

MDPs can be used to formalize a discrete-time planning task of a single agent in a stochastically changing environment, on the condition that the agent can observe the state of the environment. This is illustrated in Figure 2.1. Every time step the state changes stochastically, but the agent chooses an action that selects a particular transition function. Taking an action a from a particular state s_t at time step t induces a probability distribution over states s_{t+1} at time step $t+1$. The goal of planning for such an MDP is to find a *policy* that is optimal with respect to the desired behavior. This desired behavior, the agent's objective, can be formulated in several ways. The first type of objective of an agent is reaching a specific goal state, for example in a maze in which the agent's goal is to reach the exit. A different formulation is given by associating a certain cost with the execution of a particular action in a particular state, in which case the goal will be to minimize the expected total cost. Alternatively, one can associate rewards with actions performed in a certain state, the goal being to maximize the total reward.

When the agent knows the probabilities of the state transitions, i.e., when it knows the model, it can contemplate the expected transitions over time and compute a plan that is either most likely to reach a specific goal state, that minimizes the expected costs or that maximizes the expected reward.

In the finite-horizon case an agent can employ a nonstationary policy $\pi = (\delta_0, \ldots, \delta_{h-1})$, which is a sequence of *decision rules* δ_t, one for each stage t. Each decision rule maps states to actions. In the infinite-horizon case, under some assumptions [Bertsekas, 2007], an agent can employ a stationary policy $\pi = (\delta)$, which is used at each stage. As such, the task of planning can be seen as a search over the space of (sequences of) decision rules. In planning, this search uses the MDP model to compute the expected rewards realized by different candidate solutions.

Such a planning approach stands in contrast to reinforcement learning (RL) [Sutton and Barto, 1998, Wiering and van Otterlo, 2012], where the agent does *not* have a model of the environment, but has to learn good behavior by repeatedly interacting with the environment. Reinforcement learning can be seen as the combined task of learning the model of the environment *and* planning, although in practice it often is not necessary to explicitly recover the environment model. This book focuses only on planning, but considers two factors that complicate computing successful plans: the inability of the agent to observe the state of the environment as well as the presence of multiple agents.

2.1.2 POMDPs

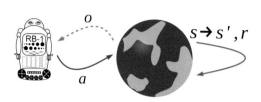

Fig. 2.2: Schematic representation of a POMDP. Instead of observing the resulting state s', the agent receives an observation $o \sim O(\cdot|s,a)$.

As mentioned in Section 1.3, noisy and limited sensors may prevent the agent from observing the state of the environment, because the observations are inaccurate and perceptual aliasing may occur. In order to represent such state uncertainty, a *partially observable Markov decision process (POMDP)* extends the MDP model by incorporating observations and their probability of occurrence conditional on the state of the environment [Kaelbling et al., 1998, Cassandra, 1998, Spaan, 2012]. This is illustrated in Figure 2.2. In a POMDP, an agent no longer knows the state of the world, but rather has to maintain a *belief* over states. That is, it can use the history of observations to estimate the probability of each state and use this information to decide upon an action.

Definition 1 (Belief). A *belief* of an agent in a POMDP is a probability distribution over states:

$$\forall_{s_t} \quad b(s_t) \triangleq \Pr(s_t|o_t,a_{t-1},o_{t-1},\ldots,a_1,o_1,a_0). \tag{2.1.1}$$

As a result, a single agent in a partially observable environment can specify its policy as a series (one for each stage t) of mappings, or decision rules, from the set of beliefs to actions. Again, for infinite-horizon settings one can usually use a stationary policy, which consists of a single decision rule used for all stages. During execution, the agent can incrementally update its current belief by using Bayes' rule. The updated belief for a particular action a_t taken and observation o_{t+1} received is given by:

$$\forall_{s_{t+1}} \quad b_{t+1}(s_{t+1}) = \frac{1}{\Pr(o_{t+1}|b_t,a_t)} \sum_{s_t} b_t(s_t)\Pr(s_{t+1},o_{t+1}|s_t,a_t). \tag{2.1.2}$$

In this equation, $\Pr(o_{t+1}|a_t,b_t)$ is a normalizing constant, and $\Pr(s_{t+1},o_{t+1}|s_t,a_t)$ is the probability that the POMDP model specifies for receiving the particular new state s_{t+1} and the resulting observation o_{t+1} assuming s_t was the previous state.

In control theory, the (continuous) observations, also referred to as measurements, are typically described as a deterministic function of the state. Sensor noise is modeled by adding a random disturbance term to this function and is dealt with by introducing a state estimator component, e.g., by Kalman filtering [Kalman, 1960]. Perceptual aliasing arises when a state component cannot be measured directly. For instance, it may not be possible to directly measure angular velocity of a robot arm; in this case it may be possible to use a so-called *observer* to estimate this velocity from its positions over time.

Although the treatment of state uncertainty in classical control theory involves terminology and techniques different from those in used in POMDPs, the basic idea in both is the same: use information gathered from the history of observations in order to improve decisions. There also is one fundamental difference, however. Control theory typically separates the estimation from the control component. For example, the estimator returns a particular value for the angles and angle velocities of the robot arm and these values are used to select actions *as if there was no uncertainty*. In contrast, POMDPs allow the agent to *explicitly reason over the belief* and what the best action is given that belief. As a result, agents using POMDP techniques can reason about information gathering: when beneficial, they will select actions that will provide information about the state.

2.2 Multiagent Decision Making: Decentralized POMDPs

Although POMDPs provide principled treatment of state uncertainty, they only consider a single agent. In order to deal with the effects of uncertainty with respect to other agents, this book will consider an extension of the POMDP framework, called *decentralized POMDP (Dec-POMDP)*.

The Dec-POMDP framework is illustrated in Figure 2.3. As the figure shows, it generalizes the POMDP to multiple agents and thus can be used to model a team of cooperative agents that are situated in a stochastic, partially observable environment. Formally, a Dec-POMDP can be defined as follows.[1]

Definition 2 (Dec-POMDP). A *decentralized partially observable Markov decision process* is defined as a tuple $\mathcal{M}_{DecP} = \langle \mathbb{D}, \mathbb{S}, \mathbb{A}, T, \mathbb{O}, O, R, h, b_0 \rangle$, where

- $\mathbb{D} = \{1,\dots,n\}$ is the set of n agents.
- $\mathbb{S}$ is a (finite) set of states.
- $\mathbb{A}$ is the set of joint actions.
- T is the transition probability function.

---

[1] Pynadath and Tambe [2002] introduced a model called multiagent team decision problem (MTDP), which is essentially equivalent to the Dec-POMDP.

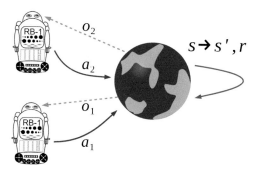

Fig. 2.3: Schematic representation of a Dec-POMDP. At every stage, each agent takes an action based on only its own observations.

- $\mathbb{O}$ is the set of joint observations.
- O is the observation probability function.
- R is the immediate reward function.
- h is the horizon of the problem as mentioned above.
- $b_0 \in \triangle(\mathbb{S})$ is the initial state distribution at time $t = 0$.[2]

The Dec-POMDP model extends single-agent POMDP models by considering *joint* actions and observations. In particular, $\mathbb{A} = \times_{i \in \mathbb{D}} \mathbb{A}_i$ is the set of *joint actions*. Here, $\mathbb{A}_i$ is the set of actions available to agent i, which can be different for each agent. At every stage t, each agent i takes an action $a_{i,t}$, leading to one joint action $a = \langle a_1, ..., a_n \rangle$ at every stage.[3] How this joint action influences the environment is described by the transition function T, which specifies $\Pr(s'|s,a)$. In a Dec-POMDP, agents only know their own individual action; they do not observe each other's actions. We will assume that $\mathbb{A}_i$ does not depend on the stage or state of the environment (but generalizations that do incorporate such constraints are straightforward to specify). Similar to the set of joint actions, $\mathbb{O} = \times_{i \in \mathbb{D}} \mathbb{O}_i$ is the set of joint observations, where $\mathbb{O}_i$ is a set of observations available to agent i. At every time step the environment emits one joint observation $o = \langle o_1, ..., o_n \rangle$ from which each agent i only observes its own component o_i. The observation function O specifies the probabilities $\Pr(o|a,s')$ of joint observations. Figure 2.4 further illustrates the dynamics of the Dec-POMDP model.

The immediate reward function $R : \mathbb{S} \times \mathbb{A} \to \mathbb{R}$ maps states and joint actions to real numbers and is used is used to specify the goal of the agents. In particular, R only specifies the *immediate* reward that is awarded for each joint action. The goal, however, should be to optimize the behavior of the team of agents over a longer term, i.e., it should optimize over all h stages. Therefore, in order to fully specify the problem, one needs to select an *optimality criterion* that indicates how the immediate rewards are combined into a single number. For instance, when planning over

---

[2] We write $\triangle(\cdot)$ for the simplex, the set of probability distributions, over the set $(\cdot)$.

[3] Note that we will write a_i for the action of agent i (when t is left unspecified) and a_t for the joint action at stage t. From the context it should be clear which is intended.

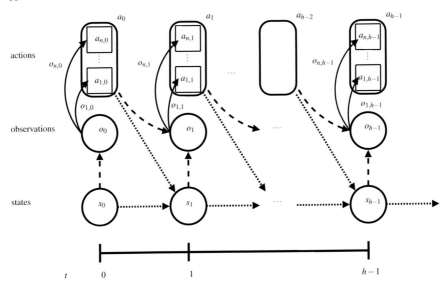

Fig. 2.4: A more detailed illustration of the dynamics of a Dec-POMDP. At every stage the environment is in a particular state. This state emits a joint observation according to the observation model (dashed arrows) from which each agent observes its individual component (indicated by solid arrows). Then each agent selects an action, together forming the joint action, which leads to a state transition according to the transition model (dotted arrows).

a finite horizon the *undiscounted expected cumulative reward* (the expectation of the sum of the rewards for all stages, introduced in Chapter 3) is commonly used as the optimality criterion. The planning problem amounts to finding a tuple of policies, called a *joint policy,* that maximizes the optimality criterion.

During execution, the agents are assumed to act based on their individual observations only and no additional communication is assumed. This does not mean that Dec-POMDPs cannot model settings which concern communication. For instance, if one agent has an action "mark blackboard" and the other agent has an observation "mark on blackboard", the agents have a mechanism of communication through the state of the environment. However, rather than making this communication explicit, we say that the Dec-POMDP can model communication implicitly through the actions, states and observations. This means that in a Dec-POMDP, communication has no special semantics. Section 8.3 further elaborates on communication in Dec-POMDPs.

Note that, as in other planning models (and in contrast to what is usual in reinforcement learning), in a Dec-POMDP, the agents are assumed not to observe the immediate rewards. Observing the immediate rewards could convey information regarding the true state, which is not present in the received observations, which is undesirable as all information available to the agents should be modeled in the observations. When planning for Dec-POMDPs, the only thing that matters is the *expectation* of the cumulative future reward, which is available in the offline planning phase, not the actual reward obtained. It is not even assumed that the actual reward can be observed at the end of the episode. If rewards are to be observed, they should be made part of the observation.

2.3 Example Domains

To illustrate the Dec-POMDP model, we discuss a number of example domains and benchmark problems. These range from the toy (but surprisingly hard) 'decentralized tiger' problem to multirobot coordination and communication network optimization.

2.3.1 Dec-Tiger

We will consider the decentralized tiger (DEC-TIGER) problem Nair et al. [2003c]—a frequently used Dec-POMDP benchmark—as an example. It concerns two agents that are standing in a hallway with two doors. Behind one door, there is a treasure and behind the other is a tiger, as illustrated in Figure 2.5.

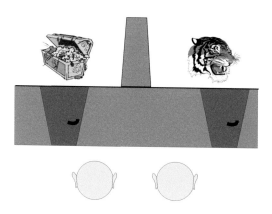

Fig. 2.5: The DEC-TIGER benchmark.

The state describes which door the tiger is behind—left (s_l) or right (s_r), each occurring with 0.5 probability (i.e., the initial state distribution b_0 is uniform). Each agent can perform three actions: open the left door (a_{OL}), open the right door (a_{OR}) or listen (a_{Li}). Clearly, opening the door to the treasure will yield a reward ($+10$), but opening the door to the tiger will result in a severe penalty (-100). A greater reward ($+20$) is given for both agents opening the correct door at the same time.

As such, a good strategy will probably involve listening first. The listen actions, however, also have a minor cost (a negative reward of -1). The full reward model is shown in Table 2.1.

a	s_l	s_r
$\langle a_{\text{Li}},a_{\text{Li}} \rangle$	-2	-2
$\langle a_{\text{Li}},a_{\text{OL}} \rangle$	-101	$+9$
$\langle a_{\text{Li}},a_{\text{OR}} \rangle$	$+9$	-101
$\langle a_{\text{OL}},a_{\text{Li}} \rangle$	-101	$+9$
$\langle a_{\text{OL}},a_{\text{OL}} \rangle$	-50	$+20$
$\langle a_{\text{OL}},a_{\text{OR}} \rangle$	-100	-100
$\langle a_{\text{OR}},a_{\text{Li}} \rangle$	$+9$	-101
$\langle a_{\text{OR}},a_{\text{OL}} \rangle$	-100	-100
$\langle a_{\text{OR}},a_{\text{OR}} \rangle$	$+20$	-50

Table 2.1: Rewards for DEC-TIGER.

At every stage the agents get an observation: they can either hear the tiger behind the left (o_{HL}) or right (o_{HR}) door, but each agent has a 15% chance of hearing it incorrectly (getting the wrong observation), which means that there is only a probability of $0.85 \cdot 0.85 = 0.72$ that both agents get the correct observation. Moreover, the observation is informative only if both agents listen; if either agent opens a door, both agents receive an uninformative (uniformly drawn) observation and the problem resets to s_l or s_r with equal probability. At this point the problem just continues, such that the agents may be able to open the door to the treasure multiple times. Also note that, since the only two observations the agents can get are o_{HL} and o_{HR}, the agents have no way of detecting that the problem has been reset: if one agent opens the door while the other listens, the other agent will not be able to tell that the door was opened. The full transition, observation and reward models are listed in Table 2.2.

a	$s_l \to s_l$	$s_l \to s_r$	$s_r \to s_r$	$s_r \to s_l$
$\langle a_{\text{Li}},a_{\text{Li}} \rangle$	1.0	0.0	1.0	0.0
otherwise	0.5	0.5	0.5	0.5

(a) Transition probabilities.

a	s_l		s_r	
	o_{HL}	o_{HR}	o_{HL}	o_{HR}
$\langle a_{\text{Li}},a_{\text{Li}} \rangle$	0.85	0.15	0.15	0.85
otherwise	0.5	0.5	0.5	0.5

(b) Individual observation probabilities.

Table 2.2: Transitions and observation model for DEC-TIGER.

2.3.2 Multirobot Coordination: Recycling and Box-Pushing

Dec-POMDPs can also be used for the high-level formalization of multirobot tasks. In fact, several benchmark problems are motivated by coordination in robotics. Here we will briefly describe two of them: RECYCLING ROBOTS and COOPERATIVE BOX PUSHING.

Recycling Robots   This is the problem described in Section 1.1 and can be represented as a Dec-POMDP in a natural way. The states, $\mathbb{S}$, consist of the different locations of each robot, their battery levels and the different amounts of trash in the cans. The actions, $\mathbb{A}_i$, for each robot consist of movements in different directions as well as decisions to pick up a trash can or recharge the battery (when in range of a can or a charging station). Recall that large trash cans can only be picked up by two agents jointly. The observations, $\mathbb{O}_i$, of each robot consist of its own battery level, its own location, the locations of other robots in sensor range and the amount of trash in cans within range. The rewards, R, could consist of a large positive value for a pair of robots emptying a large (full) trash can, a small positive value for a single robot emptying a small trash can and negative values for a robot depleting its battery or a trash can overflowing. An optimal solution is a joint policy that leads to the expected behavior (given that the rewards are properly specified). That is, it ensures that the robots cooperate to empty the large trash cans when appropriate and the small ones individually while considering battery usage.

Box Pushing   The COOPERATIVE BOX PUSHING domain was introduced by Seuken and Zilberstein [2007b] and is a larger two-robot benchmark. Also in this domain the agents are situated in a grid world, but now they have to collaborate to move boxes in this world. In particular, there are small boxes that can be moved by one agent, and big boxes that the agents have to push together. Each agent has four actions: turn left, turn right, move forward and stay, and five observations that describe the grid position in front of the agent: empty, wall, other agent, small box, large box.

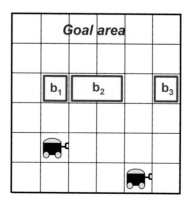

Fig. 2.6: An example cooperative box pushing problem.

2.3.3 Network Protocol Optimization

The BROADCASTCHANNEL, introduced by Hansen et al. [2004] (and modified by Bernstein et al. [2005]), models two nodes that have to cooperate to maximize the throughput of a shared communication channel. At each stage the agents can choose to send or not send a message across the channel, and they can noisily observe whether there was a collision (if both agents sent) or not. A reward of 1 is given for every stage in which a message is successfully sent over the channel and all other actions receive no reward. This problem has four states (in which each agent has a 0 or 1 message in its buffer), two actions (send or do not send) and five observations (combinations of whether there is a message in the buffer or not and whether there was a collision if a message was sent).

A more advanced version of this problem also considers state variables that are the average interarrival time of acknowledgments, the average outgoing time step from the sender's acknowledgments, and the ratio of the most recent round-trip-time measurement to the minimum observed so far. Its actions consist of adjustments to the congestion window (increments and multipliers) and the minimum interval between outgoing packets [Winstein and Balakrishnan, 2013].

2.3.4 Efficient Sensor Networks

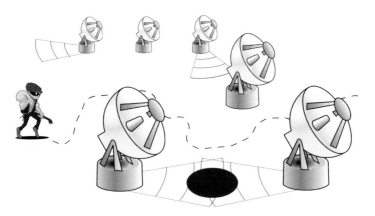

Fig. 2.7: A sensor network for intrusion detection. Scanning overlapping areas increases the chance of detection, but sensor nodes should also try to preserve power.

Sensor networks have also been modeled as Dec-POMDPs [Nair et al., 2005, Marecki et al., 2008]. For instance, consider the setting illustrated in Figure 2.7. Here, a network of sensors needs to coordinate to maximize the chance of detecting intruders, while minimizing power usage. The intruders navigate through the plane (which could be discretized as a grid) according to certain (stochastic) patterns. At every stage, sensors decide on a direction to scan or decide to turn off for that time step (thus saving energy). A typical description of the state of such a problem would include the position of the intruder, as well as any variable associated with the sensor nodes (such as remaining battery power). The observations for each sensor node would typically be whether there was a detection or not. Rewards could model the cost (energy use) of scanning, and assign a positive reward for every stage in which sensor nodes scan the area including the target, with a higher reward if multiple sensors scan that area. These problems are cooperative because the reward will often be super- (or sub)additive, reflecting the greater (or redundant) information that is gathered when multiple sensors detect a target on a single time step.

Since the numbers of states, joint actions and joint observations in a sensor network can be quite large, it is typically necessary to exploit the structure specific to these problems in order to be able to represent them compactly. This will be discussed in Section 2.4.2. The specific subclass of Dec-POMDP that is frequently used for sensor networks is called the *networked distributed POMDP (ND-POMDP)*. ND-POMDPs will be discussed in Chapter 8, where we will revisit this example.

2.4 Special Cases, Generalizations and Related Models

Because solving Dec-POMDPs is complex (as will be discussed in the next chapter), much research has focused on special cases of Dec-POMDPs. This section briefly treats a number of special cases that have received considerable attention. For a more comprehensive overview of all the special cases, the reader is referred to the articles by Pynadath and Tambe [2002], Goldman and Zilberstein [2004] and Seuken and Zilberstein [2008]. Additionally, we give a description of the partially observable stochastic game, which generalizes the Dec-POMDP, and the interactive POMDP, which is a related framework but takes a subjective perspective.

2.4.1 Observability and Dec-MDPs

One of the dimensions in which Dec-POMDPs can be distinguished is the amount of information that they provide to the agents via the observations. Research has identified different categories of observation functions corresponding to degrees of observability [Pynadath and Tambe, 2002, Goldman and Zilberstein, 2004]. When the observation function is such that the individual observation for each of the agents will always uniquely identify the true state, the problem is considered *fully observable*, also called *individually observable*. In such a case, the problem reduces to a centralized model; these will be treated in some more detail in Section 2.4.3.

The other extreme is when the problem is *non-observable,* meaning that none of the agents observes any useful information. This is modeled by restricting the set of observations to a single null-observation, $\forall_i \, \mathbb{O}_i = \{o_{i,\emptyset}\}$. Under non-observability agents can only employ an *open-loop plan*: a predetermined sequence of actions. A result of this is that the non-observable setting is easier from a complexity point of view (NP-complete, Pynadath and Tambe 2002).

Between these two extremes there are partially observable problems which are the focus of this book. One more special case has been identified, namely the case where not the individual, but the joint observation identifies the true state. In other words, if the observations of all the agents are combined, the state of the environment is known exactly. This case is referred to as *jointly* or *collectively observable*.

Definition 3 (Dec-MDP). A jointly observable Dec-POMDP is referred to as a *decentralized Markov decision process (Dec-MDP)*.

A common example of a Dec-MDP is a problem in which the state consists of the locations of a set of robots and each agent observes its own location perfectly. Therefore, if all these observations are combined, the locations of all robots would be known.

It is important to keep in mind that, even though all observations together identify the state in a Dec-MDP, each agent still has a partial view. As such, Dec-MDPs are a nontrivial subclass of Dec-POMDPs and in fact it can be shown that the worst case complexity (cf. Section 3.5) of this subclass is the same as that of the entire class of Dec-POMDPs [Bernstein et al., 2002]. This implies that hardness comes from being distributed, not (only) from having a hidden state.

2.4.2 Factored Models

A different family of special cases focuses on using properties that the transition, observation and reward function might exhibit in order to both *compactly represent* and *efficiently solve* Dec-POMDP problems. The core idea is to consider the states and transition, observation and reward functions not as atomic entities, but as consisting of a number of *factors*, and explicitly representing how different factors affect each other.

For instance, in the case of a sensor network, the observations of each sensor typically depend only on its local environment. Therefore, it can be possible to represent the observation model more compactly as a product of smaller observation functions, one for each agent. In addition, since in many cases the sensing costs are local and sensors do not influence their environment there is likely special structure in the reward and transition function.

A large number of models that exploit factorization have been proposed, such as transition- and observation-independent Dec-MDPs [Becker et al., 2003], ND-POMDPs [Nair et al., 2005], factored Dec-POMDPs [Oliehoek et al., 2008c], and many others [Becker et al., 2004a, 2005, Shen et al., 2006, Spaan and Melo, 2008, Varakantham et al., 2009, Mostafa and Lesser, 2009, Witwicki and Durfee,

2009, 2010b, Mostafa and Lesser, 2011a,b, Witwicki et al., 2012]. Some of these will be treated in more detail in Chapter 8. In the remainder of this section, we give an overview of a number of different forms of independence that can arise.

We will discuss factorization in the context of Dec-MDPs, but similar factorization can be done in full Dec-POMDPs.

Definition 4 (Agent-Wise Factored Dec-MDP). An *(agent-wise) factored n-agent Dec-MDP* is a Dec-MDP where the world state can be factored into $n + 1$ components, $\mathbb{S} = \mathbb{S}_1 \times \ldots \times \mathbb{S}_n$. The states in $s_i \in \mathbb{S}_i$ are the *local states* associated with agent i. [4]

For example, consider an agent navigation task where the agents are located in positions in a grid and the goal is for all agents to navigate to a particular grid cell. In such a task, an agent's local state, s_i, might consist of its location in a grid. Next, we identify some properties that an agent-wise factored Dec-MDP might posses.

An agent-wise factored Dec-MDP is said to be *locally fully observable* if each agent fully observes its own state component. For instance, if each agent in the navigation problem can observe its own location the state is locally fully observable.

A factored, n-agent Dec-MDP is said to be *transition-independent* if the state transition probabilities factorize as follows:

$$T(s' \mid s,a) = \prod_{i}^{n} T_i(s_i' \mid s_i,a_i).$$ (2.4.1)

Here, $T_i(s_i' \mid s_i,a_i)$ represents the probability that the local state of agent i transitions from s_i to s_i' after executing action a_i. For instance, a robot navigation task is transition-independent if the robots never affect each other (i.e., they do not bump into each other when moving and can share the same grid cell). On the other hand, RECYCLING ROBOTS (see Section 2.3.2) is not transition-independent. Even though the movements are independent, the state cannot be factored into local components for each agent: this would require an arbitrary assignment of small trash cans to agents; moreover, no agent can deal with the large trash cans by itself.

A factored, n-agent Dec-MDP is said to be *observation-independent* if the observation probabilities factorize as follows:

$$O(o \mid a,s') = \prod_{i \in \mathbb{D}} O_i(o_i \mid a_i,s_i').$$ (2.4.2)

In the equation above, $O_i(o_i \mid a_i,s_i')$ represents the probability that agent i receives observation o_i in state s_i' after executing action a_i. If the robots in the navigation problem cannot observe each other (due to working in different locations or lack of sensors), the problem becomes observation-independent.

A factored, n-agent Dec-MDP is said to be *reward-independent* if there is a monotonically nondecreasing function f such that

---

[4] Some factored models also consider an s_0 component that is a property of the environment and is not affected by any agent actions.

$$R(s,a) = f(R_1(s_1,a_1),\ldots,R_n(s_n,a_n)),\qquad (2.4.3)$$

If this is the case, the global reward is maximized by maximizing local rewards. For instance, additive local rewards,

$$R(s,a) = \sum_{i\in\mathbb{D}} R_i(s_i,a_i),\qquad (2.4.4)$$

are frequently used.

2.4.3 Centralized Models: MMDPs and MPOMDPs

In the discussion so far we have focused on models that, in the execution phase, are truly decentralized: they model agents that select actions based on local observations. A different approach is to consider models that are *centralized*, i.e., in which (joint) actions can be selected based on global information. Such global information can arise due to either full observability or communication. In the former case, each agent simply observes the same observation or state. In the latter case, we have to assume that agents can share their individual observations over an instantaneous and noise-free communication channel without costs. In either case, this allows the construction of a centralized model.

For instance, under such communication, a Dec-MDP effectively reduces to a *multiagent Markov decision process (MMDP)* introduced by Boutilier [1996].

Definition 5 (MMDP). A *multiagent Markov decision process (MMDP)* is defined as a tuple $\mathcal{M}_{MMDP} = \langle \mathbb{D}, \mathbb{S}, \mathbb{A}, T, R, h \rangle$, where the components are identical to the case of Dec-POMDPs (see Definition 2).

In this setting a joint action can be selected based on the state without considering the history, because the state is Markovian and known by all agents. Moreover, because each agent knows what the state is, there is an effective way to coordinate. One can think of the situation as a regular MDP with a 'puppeteer' agent that selects joint actions. For this 'underlying MDP' an optimal solution π^* can be found efficiently[5] with standard dynamic programming techniques [Puterman, 1994]. Such a solution $\pi^* = (\delta_0,\ldots,\delta_{h-1})$ specifies a mapping from states to joint actions for each stage $\forall_t\ \delta_t : \mathbb{S} \to \mathbb{A}$ and can be split into individual policies $\pi_i = (\delta_{i,0},\ldots,\delta_{i,h-1})$ with $\forall_t\ \delta_{i,t} : \mathbb{S} \to \mathbb{A}_i$ for all agents.

Similarly, adding broadcast communication to regular Dec-POMDP results in a *multiagent POMDP (MPOMDP)*, which is a special type of POMDP. In this MPOMDP, each agent can compute the *joint belief:* i.e., the probability distribution over states given the histories of joint actions and observations.

---

[5] Solving an MDP is P-complete [Papadimitriou and Tsitsiklis, 1987], but the underlying MDP of a Dec-POMDP still has size exponential in the number of agents. However, given the MMDP representation for a particular (typically small) number of agents, the solution is efficient.

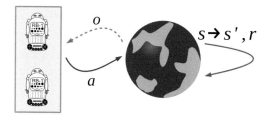

Fig. 2.8: Schematic representation of an MPOMDP. At every stage, each agent broadcasts its individual observation, which means that effectively each agent receives the joint observation.

Definition 6 (Joint Belief). A *joint belief* is the probability distribution over states induced by the initial state distribution b_0 and the history of joint actions and observations:

$$b(s) \triangleq \Pr(s|b_0,a_o,o_1,a_1,\dots,a_{t-1}\,o_t).$$

We will also write $\mathbb{B} = \triangle(\mathbb{S})$ for the set of joint beliefs.

Since the MPOMDP is a POMDP, the computation of this joint belief can be done incrementally using Bayes' rule in exactly the same way as described in Section 2.1.2.

Even though MPOMDPs are POMDPs and POMDPs are intractable to solve (PSPACE-complete, Papadimitriou and Tsitsiklis 1987), solving an MPOMDP is usually easier than solving a Dec-POMDP in practice. The solution of an MPOMDP specifies a mapping from joint beliefs to joint actions for each stage, $\forall_t \; \delta_t :$ $\triangle(\mathbb{S}) \to \mathbb{A}$, and can be split into individual policies $\pi_i = (\delta_{i,0},\dots,\delta_{i,h-1})$ with $\forall_t \; \delta_{i,t} : \triangle(\mathbb{S}) \to \mathbb{A}_i$ for all agents.

2.4.4 Multiagent Decision Problems

The attentive reader might wonder why we have not given a definition in terms of a formal tuple for the MPOMDP framework. The reason is that this definition would be identical to the definition of the Dec-POMDP given in Definition 2. That is, the traditional definition of a Dec-POMDP presented in Section 2.2 is underspecified since it does not include the specification of the communication capabilities of the agents. We try and rectify this situation here.

In particular, we introduce a formalization of a more general class of *multiagent decision problems (MADPs)* that will make more explicit all the constraints specified by its members. In particular, it will make clearer what the decentralization constraints are that the Dec-POMDP model imposes, and how the approach can be generalized (e.g., to deal with different assumptions with respect to communication). We begin by defining the *environment* of the agents:

Definition 7 (Markov Multiagent Environment). The *Markov multiagent environment (MME)* is defined as a tuple $\mathcal{M} = \langle \mathbb{D}, \mathbb{S}, \mathbb{A}, T, \mathbb{O}, O, \mathbb{R}, h, b_0 \rangle$, where

- $\mathbb{R} = \{R_1, \dots R_n\}$ is the *set* of immediate reward functions for all agents,
- all the other components are exactly the same as in our previous definition of Dec-POMDP (Definition 2).

With the exception of the remaining two subsections, in this book we restrict ourselves to collaborative models: a *collaborative MME* is an MME where all the agents get the same reward:

$$\forall_{i,j} \qquad R_i(s,a) = R_j(s,a),$$

which will be simply written as $R(s,a)$.

An MME is underspecified in that it does not specify the information on which the agents can base their actions, or how they update their information. We make this explicit by defining an agent model.

Definition 8 (Agent Model). A *model* for agent i is a tuple $m_i = \langle \mathbb{I}_i, I_i, \mathbb{A}_i, \mathbb{O}_i, \mathbb{Z}_i, \pi_i, \iota_i \rangle$, where

- $\mathbb{I}_i$ is the set of *information states (ISs)* (also *internal states,* or *beliefs*),
- I_i is the *current* internal state of the agent,
- $\mathbb{A}_i, \mathbb{O}_i$ are as before: the actions taken by / observations that the environment provides to agent i,
- $\mathbb{Z}_i$ is the set of *auxiliary observations z_i* (e.g., from communication) available to agent i,
- π_i is a (stochastic) action selection policy $\pi_i : \mathbb{I}_i \to \triangle(\mathbb{A}_i)$,
- ι_i is the (stochastic) *information state function* (or *belief update function*) $\iota_i : \mathbb{I}_i \times \mathbb{A}_i \times \mathbb{O}_i \times \mathbb{Z}_i \to \triangle(\mathbb{I}_i)$.

This definition makes clear that the MME framework leaves the specification of the auxiliary observations, information states, the information state function, as well as the action selection policy unspecified. As such, the MME by itself is not enough to specify a dynamical process. Instead, it is necessary to specify those missing components for all agents. This is illustrated in Figure 2.9, which shows how a dynamic multiagent system (in this case, a Dec-POMDP, which we redefine below) evolves over time. It makes clear that there is a environment component, the MME, as well as an *agent component* that specifies how the agents update their internal state, which in turn dictates their actions.[6] It is only these two components together that lead to a dynamical process.

Definition 9 (Agent Component). A *fully specified agent component,* can be formalized as a tuple $m = \langle \mathbb{D}, \{\mathbb{I}_i\}, \{I_{i,0}\}, \{\mathbb{A}_i\}, \{\mathbb{O}_i\}, \{\mathbb{Z}_i\}, \{\iota_i\}, \{\pi_i\} \rangle$, where

- $\mathbb{D} = \{1, \dots, n\}$ is the set of n agents.

---

[6] In the most general form, the next internal states would explicitly depend on the taken action too (not shown, to avoid clutter).

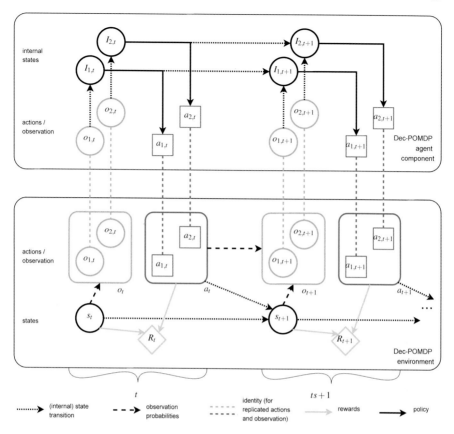

Fig. 2.9: Illustration of the new perspective on the Dec-POMDP for the two-agent case. The process is formed by an environment and an agent component that together generate the interaction over time.

- $\{\mathbb{I}_i\}$ are the sets of internal states for each agent.
- $\{I_{i,0}\}$ are the initial internal states of each agent.
- $\{\mathbb{A}_i\}$ are the sets of actions.
- $\{\mathbb{O}_i\}$ are the sets of observations.
- $\{\mathbb{Z}_i\}$ are the sets of auxiliary observations.
- $\{\iota_i\}$ are the information state functions for each agent.
- $\{\pi_i\}$ are the policies, that map from internal states to actions.

Additionally, we assume that the agent component specifies a mechanism (left implicit here) for generating the auxiliary observations.

As such, the agent component handles the specification of the entire team of agents and their internal workings. That is, one can alternatively think of the agent component as a set of agent models, *a joint agent model,* for stage $t = 0$, together with a mechanism to generate the auxiliary observations.

Clearly, once the MME and a fully specified agent component are brought together, we have a dynamical system: a somewhat more complicated Markov reward process. The goal in formalizing these components, however, is that we want to *optimize* the behavior of the overall system. That is, we want to optimize the agent component in such a way that the reward is maximized.

As such, we provide a perspective of a whole range of *multiagent decision problems* that can be formalized in this fashion. On the one hand, the *problem designer* 1) selects an optimality criterion, 2) specifies the MME, and 3) may specify a subset of the elements of the agent component (which determines the 'type' of problem that we are dealing with). On the other hand, the *problem optimizer* (e.g., a planning method we develop) has as its goal to optimize the nonspecified elements of the agent component in order to maximize the value as given by the optimality criterion. In other words, we can think of a multiagent decision problem as the specification of an MME together with a non-fully specified agent component.

Redefining Dec-POMDPs   We can now redefine what a Dec-POMDP is by making use of this framework of MADPs.

Definition 10 (Dec-POMDP).   A *decentralized POMDP (Dec-POMDP)* is a tuple $\mathcal{M}_{DecP} = \langle OC, \mathcal{M}, m \rangle$, where

- OC is the optimality criterion,
- $\mathcal{M}$ is an MME, and
- $m = \langle \mathbb{D}, \cdot, \cdot, \{\mathbb{A}_i\}, \{\mathbb{O}_i\}, \{\mathbb{Z}_i = \emptyset\}, \cdot, \cdot \rangle$ is a partially specified agent component: m can be seen to partially specify the model for each agent: for each model m_i contained in the agent component, it specifies that $\mathbb{Z}_i = \emptyset$. That is, there are no auxiliary observations, such that each agent can form its internal state, and thus act, based only on its local actions and observations.

The goal for the problem optimizer for a Dec-POMDP is to specify the elements of m that are not specified: $\{\mathbb{I}_i\}, \{I_{i,0}\}, \{t_i\}, \{\pi_i\}$. That is, the action selection policies need to be optimized and choices need to be made with respect to the representation and updating of information states. As we will cover in more detail in later chapters, these choices are typically made differently in the finite and infinite horizon case: internal states are often represented as nodes in a tree (in the former case) or as a finite-state controller (in the latter case) for each agent.[7]

Defining MPOMDPs   Now, we can also give a more formal definition of an MPOMDP. As we indicated at the start of this section, an MPOMDP cannot be discriminated from a Dec-POMDP on the basis of what we now call the MME. Instead, it differs from a Dec-POMDP only in the partial specification of the agent component. This is illustrated in Figure 2.10. In particular, the set of internal states

---

[7] The optimality criterion selected by the problem designer also is typically different depending on the horizon: maximizing the undiscounted and discounted sum of cumulative rewards are typically considered as the optimality criteria for the finite and infinite horizon cases, respectively. While this does not change the task of the problem optimizer, it can change the methods that can be employed to perform this optimization.

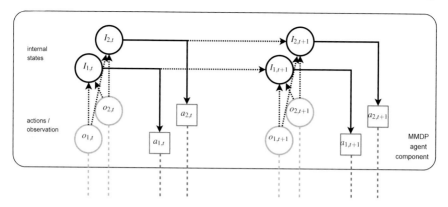

Fig. 2.10: The agent component for an MPOMDP. The agents share their individual observations via communication and therefore can maintain the same internal state. In particular, given the history of joint actions and joint observations, each agent can compute the joint belief $I_{1,t} = I_{2,t} = b_t$.

of the agents is the set of joint beliefs. This allows us to give a formal definition of the MPOMDP:

Definition 11 (MPOMDP). A *multiagent POMDP (MPOMDP)* is specified by a tuple $\mathcal{M}_{MPOMDP} = \langle OC, \mathcal{M}, m \rangle$, where

- OC is the optimality criterion,
- $\mathcal{M}$ is an MME, and
- $m = \langle \mathbb{D}, \{\mathbb{I}_i\}, \{I_{i,0}\}, \{\mathbb{A}_i\}, \{\mathbb{O}_i\}, \{\mathbb{Z}_i\}, \{\iota_i\}, \cdot \rangle$ is a partially specified agent component. For each agent i:

  - The set of internal states is the set of joint beliefs $\mathbb{I}_i = \mathbb{B}$.
  - $I_{i,0} = b_0$.
  - The auxiliary observations are the observations of the other agents $o_{-i} = \langle o_1, \ldots, o_{i-1}, o_{i+1}, \ldots, o_n \rangle$ obtained through instantaneous communication. That is, $\mathbb{Z}_i = \bigotimes_{j \neq i} \mathbb{O}_j$.
  - The information state function is specified by the joint belief update: $b_{t+1} = \iota_i(b_t, a_t, o_{t+1})$ if and only if b_{t+1} is the result of performing the belief update for a_t, o_{t+1}—cf. (2.1.2)—on b_t.

We see that in an MPOMDP many more elements of the agent component are specified. In particular, only the action selection policies $\{\pi_i\}$ that map from internal states (i.e., joint beliefs) to individual actions need to be specified.

2.4.5 Partially Observable Stochastic Games

The Dec-POMDP is a very general model in that it deals with many types of uncertainty and multiple agents. However, it is only applicable to cooperative teams of agents, since it only specifies a single (team) reward. The generalization of the Dec-POMDP is the *partially observable stochastic game (POSG)*. It has the same components as a Dec-POMDP, except that it specifies not a single reward function, but a collection of reward functions, one for each agent. This means that a POSG assumes self-interested agents that want to maximize their individual expected cumulative reward.

The consequence of this is that we arrive in the field of game theory: there is no longer an optimal joint policy, simply because optimality is no longer defined. Rather the joint policy should be a (Bayesian) Nash Equilibrium, and preferably a Pareto optimal one.[8] However, there is no clear way to identify the best one. Moreover, such a Pareto optimal NE is only guaranteed to exist in randomized policies (for a finite POSG), which means that it is no longer possible to perform brute-force policy evaluation (see Section 3.4). Also search methods based on alternating maximization (see Section 5.2.1) are no longer guaranteed to converge for POSGs. The dynamic programming method proposed by Hansen et al. [2004], covered in Section 4.1.2, does apply to POSGs: it finds the set of nondominated policies for each agent.

Even though the consequences of switching to self-interested agents are severe from a computational perspective, from a modeling perspective the Dec-POMDP and POSG framework are very similar. In particular all dynamics with respect to transitions and observations are identical, and therefore computation of probabilities of action-observation histories and joint beliefs transfers to the POSG setting. As such, even though solution methods presented in this book may not transfer directly to the POSG case, the modeling aspect largely does. For instance, the conversion of a Dec-POMDP to a type of centralized model (covered in Section 4.3) can be transferred to the POSG setting [Wiggers et al., 2015].

2.4.6 Interactive POMDPs

Both the Dec-POMDP and POSG frameworks present an *objective perspective* of the MAS: they present a picture of the whole situation and solution methods try to find plans for all agents at the same time. An alternative approach to MASs is to consider it from the perspective of a one particular agent, which we refer to as the *subjective perspective* of an MAS.

---

[8] Explanations of these concepts as well as other concepts in this section can be found in for example the texts by Binmore [1992], Osborne and Rubinstein [1994] and Leyton-Brown and Shoham [2008].

The simplest approach is to try and model the decision making process of the protagonist agent as a POMDP by simply ignoring other agents, and treating their influence on the transitions and observations as noise. This approximation has as a drawback that it decreases the value of the optimal policy. Moreover, it cannot deal with nonstationarity of the influence of other agents; in many settings the behavior of other agents can change over time (e.g., as the result of changes to their beliefs).

A more sophisticated approach is to have the protagonist agent maintain explicit models of the other agents in order to better predict them. This is the approach chosen in the *recursive modeling method (RMM)* [Gmytrasiewicz and Durfee, 1995, Gmytrasiewicz et al., 1998], which presents a stateless game framework, and the *interactive POMDP (I-POMDP)* framework [Gmytrasiewicz and Doshi, 2005], which extends this approach to sequential decision problems with states and observations.

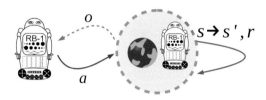

Fig. 2.11: Schematic representation of an I-POMDP. The agent reasons about the joint state of the environment and the other agent(s).

The general idea is illustrated in Figure 2.11: the protagonist agent models the *interactive state* $\check{s}_i = \langle s, m_j \rangle$ of its environment as consisting of a world state s and a model for (including the internal state of) the other agent m_j.[9] Since this interactive state is the hidden state of a POMDP, it allows the agent to deal with partial observability of the environment as well as with uncertainty regarding the model of the other agent. During execution the agent maintains an *interactive belief* over world states and models of other agents.

Definition 12. Formally, an *interactive POMDP (I-POMDP)* of agent i is a tuple $\langle \check{\mathbb{S}}_i, \mathbb{A}, T_i, R_i, \mathbb{O}_i, O_i, h \rangle$, where:

- $\check{\mathbb{S}}_i$ is the set of *interactive states*.
- $\mathbb{A}$ is the set of joint actions.
- $T_i, R_i, \mathbb{O}_i, O_i$ are the transition and reward functions, observations and the observation function for agent i. These are defined over joint actions (specifying $\Pr(s'|s,a)$ and $R_i(\check{s}_i,a)$), but over individual observations (i.e., O_i specifies $\Pr(o_i|a,s')$).

Since an I-POMDP can be seen as a POMDP defined over interactive states, the POMDP belief update can be generalized to the I-POMDP setting [Gmytrasiewicz

---

[9] This generalizes to more than two agents, but for simplicity we focus on a two-agent (i and j) setting.

and Doshi, 2005]. The intuition is that, in order to predict the actions of the other agents, it uses probabilities $\forall_j \Pr(a_j|\theta_j)$ given by the model m_j.

An interesting case occurs when considering so-called *intentional models:* i.e., when assuming the other agent also uses an I-POMDP. In this case, the formal definition of I-POMDPs as above leads to an infinite hierarchy of beliefs, because an I-POMDP for agent i defines its belief over models and thus types of other agents, which in turn define a belief over the type of agent i, etc. In response to this phenomenon, Gmytrasiewicz and Doshi [2005] define *finitely nested I-POMDPs*. Here, a 0th-level belief for agent i, $b_{i,0}$, is a belief over world states $\mathbb{S}$. An kth-level belief $b_{i,k}$ is defined over world states and models consisting of types that admit beliefs of (up to) level $k-1$. The actual number of levels that the finitely nested I-POMDP admits is called the *strategy level*.

Chapter 3
Finite-Horizon Dec-POMDPs

In this chapter, we discuss issues that are specific to finite-horizon Dec-POMDPs. First, we formalize the goal of planning for Dec-POMDPs by introducing optimality criteria and policy representations that are applicable in the finite-horizon case. Subsequently, we discuss the concepts of multiagent belief and value functions for joint policies that are crucial for many planning methods. Finally, we end this chapter with a discussion of the computational complexity of Dec-POMDPs.

3.1 Optimality Criteria

An optimality criterion defines exactly what we (i.e., the problem optimizer from Section 2.4.4) want to optimize. In particular, a desirable sequence of joint actions should correspond to a high 'long-term' reward, formalized as the *return*.

Definition 13. Let the *return* or *cumulative reward* of a Dec-POMDP be defined as the total sum of the rewards the team of agents receives during execution:

$$CR(s_0,a_0,s_1,\ldots,s_{h-1},a_{h-1}) = \sum_{t=0}^{h-1} R(s_t,a_t) \qquad (3.1.1)$$

where $R(s_t,a_t)$ is the reward received at time step t.

A very typical optimality criterion in finite horizon settings is the expectation of the CR, i.e., the *expected cumulative reward*,

$$ECR = \mathbf{E}\left[\sum_{t=0}^{h-1} R(s_t,a_t)\right], \qquad (3.1.2)$$

where the expectation refers to the expectation over sequences of states and executed joint actions. The planning problem is to find a conditional plan, or *policy*, for each agent to maximize the optimality criterion. Because the rewards depend on the

© The Author(s) 2016
F.A. Oliehoek and C. Amato, *A Concise Introduction to Decentralized POMDPs*,
SpringerBriefs in Intelligent Systems, DOI 10.1007/978-3-319-28929-8_3

actions of all agents in Dec-POMDPs, this amounts to finding a tuple of policies, called a *joint policy*, that maximizes the expected cumulative reward for the team.

Another frequently used optimality criterion is the discounted expected cumulative reward

$$DECR = \mathbf{E}\left[\sum_{t=0}^{h-1} \gamma^t R(s_t, a_t)\right], \tag{3.1.3}$$

where $0 \leq \gamma < 1$ is the discount factor. Discounting gives higher priority to rewards that are obtained sooner, which can be desirable in some applications. This can be thought of in financial terms, where money now is worth more than money received in the future. Discounting is also used to keep the optimality criterion bounded in infinite horizon problems, as we discuss in Chapter 6. Note that the regular (undiscounted) expected cumulative reward is the special case with $\gamma = 1$.

3.2 Policy Representations: Histories and Policies

In an MDP, the agent uses a policy that maps states to actions. In selecting its action, an agent can ignore the history (of states) because of the Markov property. In a POMDP, the agent can no longer observe the state, but it can compute a belief b that summarizes the history; it is also a Markovian signal. In a Dec-POMDP, however, during execution each agent will only have access to its *individual* actions and observations and there is no method known to summarize this individual history. It is not possible to maintain and update an individual belief in the same way as in a POMDP, because the transition and observation function are specified in terms of joint actions and observations.

In a Dec-POMDP, *the agents do not have access to a Markovian signal during execution.* As such, there is no known statistic into which the problem optimizer can compress the histories of actions and observations without sacrificing optimality.[1] As a consequence, planning for Dec-POMDPs involves searching the space *joint* Dec-POMDP policies that map full-length individual histories to actions. We will see later that this also means that solving Dec-POMDPs is even harder than solving POMDPs.

3.2.1 Histories

First, we define histories that are used in Dec-POMDPs.

---

[1] When assuming slightly more information *during planning*, one approach is known to compress the space of internal states: Oliehoek et al. [2013a] present an approach to lossless clustering of individual histories. This, however, does not fundamentally change the representation of all the internal states (as is done when, for example, computing a belief for a POMDP); instead only some histories that satisfy a particular criterion are clustered together.

Definition 14 (Action-Observation History). An *action-observation history (AOH) for agent i, $\bar{\theta}_i$,* is the sequence of actions taken by and observations received by agent *i*. At a specific time step *t*, this is

$$\bar{\theta}_{i,t} = (a_{i,0},o_{i,1},\ldots,a_{i,t-1},o_{i,t}). \tag{3.2.1}$$

The *joint action-observation history* $\bar{\theta}_t = \langle \bar{\theta}_{1,t},\ldots,\bar{\theta}_{n,t} \rangle$ specifies the AOH for all agents. Agent *i*'s set of possible AOHs at time *t* is $\bar{\Theta}_{i,t}$. The set of AOHs possible for all stages for agent *i* is $\bar{\Theta}_i$ and $\bar{\theta}_i$ denotes an AOH from this set.[2] Finally the set of all possible *joint* AOHs $\bar{\theta}$ is denoted $\bar{\Theta}$. At $t=0$, the (joint) AOH is empty $\bar{\theta}_0 = ()$.

Definition 15 (Observation History). An *observation history (OH) for agent i, $\bar{o}_i$,* is defined as the sequence of observations an agent has received. At a specific time step *t*, this is:

$$\bar{o}_{i,t} = (o_{i,1},\ldots,o_{i,t}). \tag{3.2.2}$$

The *joint observation history,* is the OH for all agents: $\bar{o}_t = \langle \bar{o}_{1,t},\ldots,\bar{o}_{n,t} \rangle$. The set of observation histories for agent *i* at time *t* is denoted by $\bar{\mathbb{O}}_{i,t}$. Similarly to the notation for action-observation histories, we also use $\bar{o}_i \in \bar{\mathbb{O}}_i$ and $\bar{o} \in \bar{\mathbb{O}}$.

Definition 16 (Action History). The *action history (AH) for agent i, $\bar{a}_i$,* is the sequence of actions an agent has performed:

$$\bar{a}_{i,t} = (a_{i,0},a_{i,1},\ldots,a_{i,t-1}). \tag{3.2.3}$$

Notation for joint action histories and sets are analogous to those for observation histories. Finally we note that, clearly, a (joint) AOH consists of a (joint) action and a (joint) observation history: $\bar{\theta}_t = \langle \bar{o}_t,\bar{a}_t \rangle$.

3.2.2 Policies

A policy π_i for an agent *i* maps from histories to actions. In the general case, these histories are AOHs, since they contain all information an agent has. The number of AOHs grows exponentially with the horizon of the problem: At time step *t*, there are $(|\mathbb{A}_i| \cdot |\mathbb{O}_i|)^t$ possible AOHs for agent *i*. A policy π_i assigns an action to each of these histories. As a result, the number of possible policies π_i is doubly exponential in the horizon.

Under a deterministic policy, only a subset of possible action-observation histories can be reached. This is illustrated by the left side of Figure 3.1, where the actions selected by the policy are given as gray arrows and the two possible observations are given as dashed arrows. Because one action will have probability 1 of being executed while all other actions will have probability 0, policies that only differ with respect to an AOH that can never be reached result in the same behavior. Therefore,

---

[2] In a particular Dec-POMDP, it may be the case that not all of these histories can actually be realized, because of the probabilities specified by the transition and observation model.

a deterministic policy can be specified by observation histories: when an agent selects actions deterministically, it will be able to infer what action it took from only
the observation history. Using this observation history formulation, a deterministic
policy can conveniently be represented as a tree, as illustrated by the right side of
Figure 3.1.

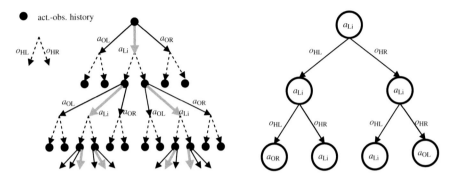

Fig. 3.1: A deterministic policy can be represented as a tree. Left: a tree of action-
observation histories $\bar{\theta}_i$ for one of the agents from the Dec-Tiger problem. A deterministic policy π_i is highlighted, showing that π_i only reaches a subset of histories
$\bar{\theta}_i$. Note that $\bar{\theta}_i$ that are not reached are not further expanded. Right: The same policy can be shown in a simplified policy tree. When both agents execute this policy
in the Dec-Tiger problem with $h = 3$, the joint policy is optimal.

Definition 17. A *deterministic policy* π_i for agent i is a mapping from observation
histories to actions, $\pi_i : \bar{\mathbb{O}}_i \to \mathbb{A}_i$.

In a deterministic policy, $\pi_i(\bar{\theta}_i)$ specifies the action for the observation his-
tory contained in (action-observation history) $\bar{\theta}_i$. For instance, if $\bar{\theta}_i = \langle \bar{o}_i, \bar{a}_i \rangle$, then
$\pi_i(\bar{\theta}_i) \triangleq \pi_i(\bar{o}_i)$. We use $\pi = \langle \pi_1, ..., \pi_n \rangle$ to denote a *joint policy*. We say that a deterministic joint policy is an *induced mapping* from joint observation histories to
joint actions $\pi : \bar{\mathbb{O}} \to \mathbb{A}$. That is, the mapping is induced by individual policies π_i
that make up the joint policy. We will simply write $\pi(\bar{o}) \triangleq \langle \pi_1(\bar{o}_1), ..., \pi_n(\bar{o}_n) \rangle$, but
note that this does not mean that π is an arbitrary function from joint observation
histories: the joint policy is *decentralized* so only a subset of possible mappings
$f : \bar{\mathbb{O}} \to \mathbb{A}$ are valid (those that specify the same individual action for each $\bar{o}_i$ of
each agent i). This is in contrast to a *centralized* joint policy that would allow any
possible mapping from *joint* histories to action (implying agents have access to the
observation histories of all other agents).

Agents can also execute stochastic policies, but (with exception of some parts
of Chapter 6 and 7) we will restrict our attention to deterministic policies without
sacrificing optimality, since a finite-horizon Dec-POMDP has at least one optimal
pure joint policy [Oliehoek et al., 2008b].

3.3 Multiagent Beliefs

As discussed in the previous section, individual agents in a Dec-POMDP cannot maintain a belief in the same way as an agent in a POMDP simply because they do not know what joint action was performed and what joint observation was emitted by the environment. Nevertheless, a number of forms of *individual belief* have been proposed in the literature [Nair et al., 2003c, Hansen et al., 2004, Oliehoek et al., 2009, Zettlemoyer et al., 2009]. In contrast to beliefs in single-agent POMDPs, these are not specified over only states, but also over histories/policies/types/beliefs of the other agents. The key point is that from an individual agent's perspective just knowing a probability distribution over states is insufficient; it also *needs to be able to predict what actions the other agents will take*.

This is illustrated most clearly by the so-called *multiagent belief*, which is a joint distribution over states and the policies that other agents will execute in the future. We postpone the formal definition of multiagent beliefs to Section 4.1.1, which formally introduces the concept of 'policies that other agents will execute in the future'.

We point out that the computation of all these types of beliefs depends, in one way or another, on the policies that have been or might be followed by some subset of agents. However, if we look at our renewed definition of a Dec-POMDP (Definition 10), we see that these quantities are not specified by the problem designer. Instead, it is up to the problem optimizer to find these. This means that all these novel notions of belief are only really useful as part of the planning process. That is, the problem optimizer can use these notions to perform 'what-if' reasoning (e.g., "what would be good individual policies if the other agents were to act in this-and-this way" or "what would be the distribution over joint histories and states if the agents follow some policy π"), but they are not useful without the problem optimizer making any such assumptions on the policies.

3.4 Value Functions for Joint Policies

Joint policies differ in how much reward they can expect to accumulate, which serves as the basis for determining their quality. Formally, we are interested in the value of the optimality criterion, the expected cumulative reward (3.1.2), that a joint policy realizes. This quantity will be simply referred to as the joint policy's *value*.

Definition 18. The *value* $V(\pi)$ of a joint policy π is defined as

$$V(\pi) \triangleq \mathbf{E}\left[\sum_{t=0}^{h-1} R(s_t, a_t) \,\Big|\, b_0, \pi\right],$$ (3.4.1)

where the expectation is over states and observations.

This expectation can be computed using a recursive formulation. For the last stage $t = h - 1$, the value is given simply by the immediate reward

$$V^\pi(s_{h-1},\bar{o}_{h-1}) = R(s_{h-1},\pi(\bar{o}_{h-1})).$$

For all other stages, the expected value is given by:

$$V^\pi(s_t,\bar{o}_t) = R(s_t,\pi(\bar{o}_t)) + \sum_{s_{t+1}\in\mathbb{S}}\sum_{o_{t+1}\in\mathbb{O}} \Pr(s_{t+1},o_{t+1}|s_t,\pi(\bar{o}_t))V^\pi(s_{t+1},\bar{o}_{t+1}).$$

$$(3.4.2)$$

Here, the probability is simply the product of the transition and observation probabilities $\Pr(s',o|s,a) = \Pr(o|a,s') \cdot \Pr(s'|s,a)$. In essence, fixing the joint policy transforms the Dec-POMDP to a Markov chain with states $(s_t,\bar{o}_t)$. Evaluating this equation via dynamic programming will result in the value for all $(s_0,\bar{o}_0)$-pairs. The value $V(\pi)$ is then given by weighting these pairs according to the initial state distribution b_0:

$$V(\pi) = \sum_{s_0\in\mathbb{S}} b_0(s_0)V^\pi(s_0,\bar{o}_0). \qquad (3.4.3)$$

(Remember $\bar{o}_0 = \langle(),\dots,()\rangle$ is the empty joint observation history, which is fixed.)

Finally, as is apparent from the above equations, the probabilities of states and histories are important in many computations. The following equation recursively specifies the probabilities of states and joint AOHs under a (potentially stochastic) joint policy:

$$\Pr(s_t,\bar{\theta}_t|b_0,\pi) = \sum_{s_{t-1}\in\mathbb{S}}\sum_{a_{t-1}\in\mathbb{A}} \Pr(s_t,o_t|s_{t-1},a_{t-1})\Pr(a_{t-1}|\bar{\theta}_{t-1},\pi)$$

$$\Pr(s_{t-1},\bar{\theta}_{t-1}|b_0,\pi). \quad (3.4.4)$$

Because there exists an optimal deterministic joint policy for a finite-horizon Dec-POMDP, it is possible to enumerate all joint policies, evaluate them as described above and choose the best one. However, the number of such joint policies is

$$O\left(|\mathbb{A}_\dagger|^{\frac{n(|\mathbb{O}_\dagger|^h - 1)}{|\mathbb{O}_\dagger| - 1}}\right), \qquad (3.4.5)$$

where $|\mathbb{A}_\dagger|$ and $|\mathbb{O}_\dagger|$ denote the largest individual action and observation sets. The cost of evaluating each joint policy is $O(|\mathbb{S}| \cdot |\mathbb{O}_\dagger|^{nh})$. It is clear that this approach therefore is only suitable for very small problems. This analysis provides some intuition about how hard the problem is. This intuition is supported by the complexity result due to Bernstein et al. [2002].

3.5 Complexity

The computational complexity of Dec-POMDPs is notorious. Intuitively, this complexity has two components: First, evaluating a policy, e.g., using (3.4.2), requires exponential time (in both the number of agents and horizon h) since we need to compute multiple values for each joint observation history. Second, and even worse, the number of such joint policies is *doubly* exponential in the horizon h.

	problem primitives				num. π for h		
	n	$|\mathbb{S}|$	$|\mathbb{A}_i|$	$|\mathbb{O}_i|$	2	4	6
DEC-TIGER	2	2	3	2	$7.29e02$	$2.06e14$	$1.31e60$
BROADCASTCHANNEL	2	4	2	2	$6.40e01$	$1.07e09$	$8.51e37$
GRIDSMALL	2	16	5	2	$1.563e04$	$9.313e20$	$1.175e88$
COOPERATIVE BOX PUSHING	2	100	4	5	$1.68e7$	$6.96e187$	$1.96e4703$
RECYCLING ROBOTS	2	4	3	2	$7.29e02$	$2.06e14$	$1.31e60$

Table 3.1: The number of joint policies for different benchmark problems and horizons.

To get an idea of what this means in practice, Table 3.1 lists the number of joint policies for a number of benchmark problems. Clearly, approaches that exhaustively search the space of joint policies have little chance of scaling beyond very small problems. Unfortunately, the complexity result due to Bernstein et al. [2002] suggests that, in the worst case, the complexity associated with such an exhaustive approach might not be avoidable.

Theorem 1 (Dec-POMDP complexity). *The problem of finding the optimal solution for a finite-horizon Dec-POMDP with $n \geq 2$ is NEXP-complete.*

Proof. The proof is by reduction from the TILING problem. See Bernstein et al. [2002] for details.

NEXP is the class of problems that in the worst case take nondeterministic exponential time. Nondeterministic means that, similarly to NP, solving these problems requires generating a guess about the solution in a nondeterministic way. Exponential time means that verifying whether the guess is a solution takes exponential time. In practice this means that (assuming NEXP $\neq$ EXP) solving a Dec-POMDP takes doubly exponential time in the worst case. Moreover, Dec-POMDPs cannot be approximated efficiently: Rabinovich et al. [2003] showed that even finding an 'ε-approximate solution' is NEXP-complete. That is, given some positive real number ε, the problem of finding a joint policy that has a value $V(\pi)$ such that $V(\pi) \geq V(\pi^*) - \varepsilon$ is also intractable. The infinite-horizon problem is undecidable, which is a direct result of the undecidability of (single-agent) POMDPs over an infinite horizon [Madani et al., 1999].

As mentioned in Section 2.4, many of the special cases are motivated by the intractability of the overall problem. However, it turns out that very strong assumptions need to be imposed in order to lower the computational complexity, as demonstrated by Table 3.2, which shows the complexity of different subclasses of Dec-MDPs. The simplest case results from having independent transitions, observations and rewards. It is straightforward to see that in this case, the problem can be decomposed into n separate MDPs and their solution can then be combined. When only the transitions and observations are independent, the problem becomes NP-complete. Intuitively, this occurs because the other agents' policies do not affect an agent's state (only the reward attained at the set of local states). Because independent transitions and observations imply local full observability, an agent's observation history does not provide any additional information about its own state—it is already known. Similarly, an agent's observation history does not provide any additional information about the other agents' states because they are independent. As a result, optimal policies become mappings from local states to actions instead of mappings from observation histories (or local state histories, as local states are locally fully observable in this case) to actions. All other combinations of independent transitions, observations and rewards do not reduce the complexity of the problem, leaving it NEXP-complete in the worst case.

Table 3.2: Complexity of finite-horizon Dec-MDP subclasses using the independence notions from Section 2.4.2.

Independence	Complexity
Transitions, observations and rewards	P-complete
Transitions and observations	NP-complete
Any other subset	NEXP-complete

Chapter 4
Exact Finite-Horizon Planning Methods

This chapter presents an overview of exact planning methods for finite-horizon Dec-POMDPs. This means that these methods perform a search through the space of joint policy trees. There are three main approaches to doing this: dynamic programming, which will be treated in Section 4.1, heuristic search, which will be treated in Section 4.2, and converting to a special case of single-agent POMDP, treated in Section 4.3. Finally, a few other methods will be treated in Section 4.4.

4.1 Backwards Approach: Dynamic Programming

In this section we treat dynamic programming for Dec-POMDPs (DP) [Hansen et al., 2004]. This method starts at the last stage, $t = h - 1$, and works its way back to the first stage, $t = 0$. As such, we say that DP works *backwards* through time, or is a *bottom-up* algorithm. At every stage t the algorithm keeps the solutions that are potentially optimal for the remaining stages $t, \ldots, h - 1$. This is similar to dynamic programming for (single-agent) MDP [Puterman, 1994, Sutton and Barto, 1998], but in contrast to that setting it will not be possible to represent these solutions using a simple value function over states. Instead, DP for Dec-POMDPs will need to maintain partial policies, called subtree policies, and values for them.

4.1.1 Growing Policies from Subtree Policies

Since policies can be represented as trees (remember Figure 3.1), a way to decompose them is by considering subtrees. Define the *time-to-go*, τ, at stage t as

$$\tau = h - t. \tag{4.1.1}$$

© The Author(s) 2016
F.A. Oliehoek and C. Amato, *A Concise Introduction to Decentralized POMDPs*, SpringerBriefs in Intelligent Systems, DOI 10.1007/978-3-319-28929-8_4

Now q_i^τ denotes a τ-stage-to-go *subtree policy* for agent i. That is, q_i^τ is a policy tree that has the same form as a full policy for the horizon-τ problem. Within the original horizon-h problem q_i^τ is a candidate for execution starting at stage $t = h - \tau$. The set of τ-stage-to-go subtree policies for agent i is denoted by $\mathbb{Q}_i^\tau$. A *joint* subtree policy $q^\tau \in \mathbb{Q}^\tau$ specifies a subtree policy for each agent: $q^\tau = \langle q_1^\tau, \ldots, q_n^\tau \rangle$.

Figure 4.1 shows different structures in a policy for a fictitious Dec-POMDP with $h = 3$. This full policy also corresponds to a 3-stage-to-go subtree policy q_i^3; two of the subtree policies are indicated using dashed ellipses.

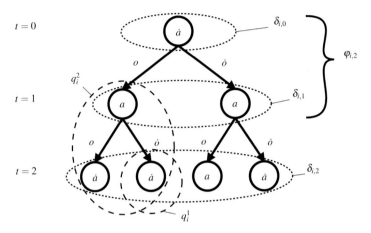

Fig. 4.1: Structure of a policy for an agent with actions $\{a, \dot{a}\}$ and observations $\{o, \dot{o}\}$. A policy π_i can be divided into *decision rules* δ_i (which are introduced in Section 4.2.1) or *subtree policies* q_i.

Subtree policies exactly correspond to the notion of 'policies that other agents will execute in the future' mentioned in Section 3.3, and allow us to formally define multiagent beliefs.

Definition 19 (Multiagent Belief [Hansen et al., 2004]). Let q_{-i}^τ be a profile of $\tau = h - t$ stage-to-go subtree policies for all agents $j \neq i$. A *multiagent belief* b_i for agent i is a joint distribution over states and subtree policies of other agents. The probability of a particular (s_t, q_{-i}^τ)-pair is written as $b_i(s_t, q_{-i}^\tau)$.

Policy trees are composed of subtrees; we refer to the reverse operation—returning the subtree for a particular observation—as policy consumption.

Definition 20 (Subtree Policy Consumption). Providing a length-τ (joint) subtree policy q^τ with a sequence of $l < \tau$ (joint) observations *consumes* a part of q^τ, leading to a (joint) subtree policy which is a subtree of q^τ. In particular, consumption $\Downarrow$ by a single joint observation o is written as

$$q^{\tau-1} = q^\tau \Downarrow_o. \tag{4.1.2}$$

For instance, in Figure 4.1, $q_i^1 = q_i^2 \Downarrow_{\bar{o}}$.

Policy consumption is important, because it allows us to reinterpret the equation of the value of a joint policy (3.4.2) in terms of subtree policies, and thus forms the basis for establishing sufficiency of the multiagent belief (to predict the value of a best response) as well as for the dynamic programming approach to Dec-POMDPs. Note that given a fixed joint policy π, a history $\bar{o}_t$ actually induces a joint subtree policy. As such, it is possible to rewrite (3.4.2) as follows. Executing q^τ over the last τ stages, starting from a state s_t at stage $t = h - \tau$ will achieve:

$$V(s_t, q^\tau) = \begin{cases} R(s_t, a_t) & \text{if } t = h - 1, \\ R(s_t, a_t) + \sum_{s_{t+1} \in \mathbb{S}} \sum_{o \in \mathbb{O}} \Pr(s_{t+1}, o | s_t, a_t) V(s_{t+1}, q^\tau \Downarrow_o) & \text{otherwise,} \end{cases}$$
(4.1.3)

where a_t is the joint action specified by (the roots of) q^τ.

This equation also explains *why* a multiagent belief $b_i(s, q_{-i}^\tau)$ is a sufficient statistic for an agent to select its policy; we can define the value of an individual q_i^τ at a multiagent belief b_i using (4.1.3):

$$V(b_i, q_i^\tau) = \sum_{s, q_{-i}^\tau} b_i(s, q_{-i}^\tau) V(s, q_{-i}^\tau, q_i^\tau).$$
(4.1.4)

This enables an agent to determine its best individual subtree policy at a multiagent belief, i.e., a multiagent belief is a sufficient statistic for agent i to optimize its policy. In addition, a multiagent belief is sufficient to predict the next multiagent belief (given a set of policies for the other agents): an agent i, after performing $a_{i,t}$ and receiving $o_{i,t+1}$, can maintain a multiagent belief via Bayes' rule. Direct substitution in (2.1.2) yields:

$$\forall_{s_{t+1}, q_{-i}^{\tau-1}} \quad b_{i,t+1}(s_{t+1}, q_{-i}^{\tau-1}) = \frac{1}{\Pr(o_{i,t+1} | b_{i,t}, a_{i,t})}$$
$$\sum_{s_t, q_{-i}^\tau} b_i(s_t, q_{-i}^\tau) \Pr(s_{t+1}, q_{-i}^{\tau-1}, o_{i,t+1} | s_t, q_{-i}^\tau, a_{i,t}),$$

where the transition and observation probabilities are the result of marginalizing over the observations that the other agents could have received:

$$\Pr(s_{t+1}, q_{-i}^{\tau-1}, o_{i,t+1} | s_t, q_{-i}^\tau, a_{i,t}) = \sum_{o_{-i,t+1}} \Pr(s_{t+1}, o_{t+1} | s_t, a_t) \mathbf{1}_{\{q_{-i}^{\tau-1} = q_{-i}^\tau \Downarrow_{o_{-i,t+1}}\}}.$$

($\mathbf{1}_{\{\cdot\}}$ is the indicator function, which is 1 if $\{\cdot\}$ is true and 0 otherwise.)

4.1.2 Dynamic Programming for Dec-POMDPs

The core idea of DP is to incrementally construct sets of longer subtree policies for the agents: starting with a set of one-stage-to-go ($\tau = 1$) subtree policies (actions) that can be executed at the last stage, construct a set of two-step policies to be executed at $h - 2$, etc. That is, DP constructs $\mathbb{Q}_i^1, \mathbb{Q}_i^2, \ldots, \mathbb{Q}_i^h$ for all agents i. When the last backup step is completed, the optimal policy can be found by evaluating all induced joint policies $\pi \in \mathbb{Q}_1^h \times \cdots \times \mathbb{Q}_n^h$ for the initial belief b_0 as described in Section 3.4.

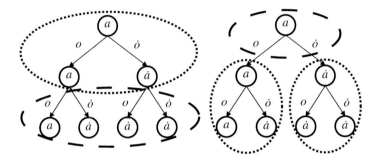

Fig. 4.2: Policy construction in MAA* (discussed in Section 4.2.2 and shown left) and dynamic programming (shown right). The figure shows how policies are constructed for an agent with two actions a, $\dot{a}$ and two observations o, $\dot{o}$. Dashed components are newly generated, dotted components result from the previous iteration.

DP formalizes this idea using *backup* operations that construct $\mathbb{Q}_i^{\tau+1}$ from $\mathbb{Q}_i^\tau$. For instance, the right side of Figure 4.2 shows how q_i^3, a three-stage-to-go subtree policy, is constructed from two $q_i^2 \in \mathbb{Q}_i^2$. In general, a one step extended policy $q_i^{\tau+1}$ is created by selecting a subtree policy for each observation and an action for the root. An exhaustive backup generates *all* possible $q_i^{\tau+1}$ that have policies from the previously generated set $q_i^\tau \in \mathbb{Q}_i^\tau$ as their subtrees. We will denote the sets of subtree policies resulting from exhaustive backup for each agent i by $\mathbb{Q}_{e,i}^{\tau+1}$.

Unfortunately, the exhaustive backup has an exponential complexity: if an agent has $|\mathbb{Q}_i^\tau|$ k-step trees, $|\mathbb{A}_i|$ actions, and $|\mathbb{O}_i|$ observations, there will be

$$|\mathbb{Q}_{e,i}^{\tau+1}| = |\mathbb{A}_i||\mathbb{Q}_i^\tau|^{|\mathbb{O}_i|}$$

$(k+1)$-step trees. This means that the sets of subtree policies maintained grow *doubly* exponentially with k. This makes sense: since the q_i^τ are essentially full policies for the horizon-k problem their number must be doubly exponentially in k.

To counter this source of intractability, it is possible to prune dominated subtree policies from $\mathbb{Q}_{e,i}^\tau$, resulting in smaller maintained sets $\mathbb{Q}_{m,i}^\tau$. As indicated by (4.1.4), the value of a q_i^τ depends on the multiagent belief. Therefore, a q_i^τ is *dominated* if it is not maximizing at any point in the multiagent belief space: the simplex over

---

variables: υ and $\{x_{q_{-i},s}\}$

maximize: υ

subject to:

$$\sum_{q_{-i},s} x_{q_{-i},s} V(s,q_i,q_{-i}) \geq \sum_{q_{-i},s} x_{q_{-i},s} V(q_i',q_{-i},s) + \upsilon \qquad \forall\, q_i'$$

$$x_{q_{-i},s} = 1 \text{ and } x_{q_{-i},s} \geq 0 \qquad \forall q_{-i},s$$

Fig. 4.3: The linear program (LP) to test for dominance. The LP determines if agent i's subtree policy q_i is dominated, by trying to find a multiagent belief point (encoded by the variables $\{x_{q_{-i},s}\}$) where the value of q_i is higher (by υ) than any other subtree policy q_i' (enforced by the constraints on the first line). If at the optimal solution υ is nonpositive, q_i is not the best subtree policy at any point in the multiagent belief space and can be pruned. The constraints on the second line simply guarantee that the variables encode a valid multiagent belief.

$\mathbb{S} \times \mathbb{Q}^{\tau}_{m,-i}$. It is possible to test for dominance by linear programming, explained in Figure 4.3. In order to perform pruning, DP must store all the values $V(s_t,q^{\tau})$. Removal of a dominated subtree policy q_i^{τ} of an agent i may cause a q_j^{τ} of another agent j to become dominated. Therefore DP iterates over agents until no further pruning is possible, a procedure known as *iterated elimination of dominated policies* [Osborne and Rubinstein, 1994].

Algorithm 4.1 Dynamic programming for Dec-POMDPs.

1: $\mathbb{Q}_i^1 \leftarrow \mathbb{A}_i, \forall i$ {Initialize with individual actions}
2: **for** $\tau = 2$ to h **do**
3:     **for** $i = 1$ to n **do**
4:         $\mathbb{Q}_i^{\tau} \leftarrow$ ExhaustiveBackup$(\mathbb{Q}_i^{\tau-1})$
5:     **end for**
6:     **while** some policies have been pruned **do**
7:         **for** $i = 1$ to n **do**
8:             $\mathbb{Q}_i^{\tau} \leftarrow$ Prune$(i, \mathbb{Q}_i^{\tau}, \mathbb{Q}_{-i,\tau})$
9:         **end for**
10:    **end while**
11: **end for**
12: $\mathbb{Q}^h \leftarrow \mathbb{Q}_1^h \times \cdots \times \mathbb{Q}_n^h$ {Construct the candidate full length policies}
13: **for** each $\pi \in \mathbb{Q}^h$ **do**
14:    $V(\pi) \leftarrow \sum_s b_0(s) V(\pi,s)$ {Evaluate them}
15: **end for**
16: **return** $\pi^* \leftarrow \arg\max_{\pi \in \mathbb{Q}^h} V(\pi)$

Algorithm 4.1 summarizes DP for Dec-POMDPs. In line 1, the individual $\tau = 1$-stage-to-go policies are initialized with the individual actions. Subsequently, the outer loop constructs one-step-longer subtree policies via exhaustive backups, and prunes the dominated subtree policies from the resulting sets. Finally, line 12 constructs all the candidate joint policies that consist of nondominated individual policies, and subsequently the best one is selected out of this set.

Note that in the main ('policy growing') loop of the algorithm, the stored values $V(s_t, q^\tau)$ are left implicit, even though they form a crucial part of the algorithm; in reality the exhaustive backup provides both one-step-longer individual policies, *as well as their values $V(s_t, q^\tau)$*, and those values are subsequently used in the pruning step. Additionally, it is worthwhile to know that this algorithm was proposed in the context of finding nondominated joint policies for POSGs. For more details, we refer the reader to the original paper by Hansen et al. [2004].

In practice, the pruning step in DP often is not able to sufficiently reduce the maintained sets to make the approach tractable for larger problems. However, the idea of point-based dynamic programming formed the basis for a heuristic method, which will be discussed in Section 5.2.2, that has achieved some empirical success in finding good policies for problems with very long horizons.

4.2 Forward Approach: Heuristic Search

In the previous section, we explained how dynamic programming works in a backward fashion: by constructing policies from the last stage and working back to the beginning. In this section we cover a method called Multiagent A* (MAA*) that is based on heuristic search. As we will see, it can be interpreted to work in a *top-down* fashion *forward* through time: starting at $t = 0$ and working to later stages.

4.2.1 Temporal Structure in Policies: Decision Rules

Policies specify actions for all stages of the Dec-POMDP. A common way to represent the temporal structure in a policy is to split it into *decision rules* δ_i that specify the policy for each stage. An individual policy is then represented as a sequence of decision rules $\pi_i = (\delta_{i,0}, \ldots, \delta_{i,h-1})$. Decision rules are indicated by dotted ellipses in Figure 4.1.

In the case of a deterministic policy, the form of the decision rule for stage t is a mapping from length-t observation histories to actions $\delta_{i,t} : \vec{\mathbb{O}}_{i,t} \to \mathbb{A}_i$. In the more general case its domain is the set of AOHs $\delta_{i,t} : \vec{\Theta}_{i,t} \to \mathbb{A}_i$. A joint decision rule $\delta_t = \langle \delta_{1,t}, \ldots, \delta_{n,t} \rangle$ specifies a decision rule for each agent.

We will also consider policies that are partially specified with respect to time. Formally, $\varphi_t = (\delta_0, \ldots, \delta_{t-1})$ denotes the *past joint policy* at stage t, which is a partial joint policy specified for stages $0, \ldots, t-1$. By appending a joint decision rule for stage t, we can 'grow' such a past joint policy.

Definition 21 (Policy Concatenation). We write

$$\varphi_{t+1} = (\delta_0, \ldots, \delta_{t-1}, \delta_t) = \langle \varphi_t \circ \delta_t \rangle \tag{4.2.1}$$

to denote policy concatenation.

Figure 4.1 shows a past policy $\varphi_{i,2}$ and illustrates how policy concatenation $\langle \varphi_{i,2} \circ \delta_{i,2} \rangle = \pi_i$ forms the full policy.

4.2.2 Multiagent A*

Multiagent A* (MAA*) is a method that uses the temporal structure in policies to search through the space of past joint policies φ_t [Szer et al., 2005]. As Figure (4.2) (left) illustrates, it works in a forward fashion: it starts with the actions to take at the first stage, then, in each iteration, a partial (i.e., past) joint policy is selected and extended with policy concatenation. This leads to a search tree where nodes correspond to past joint policies, shown in Figure 4.4: the root node is a completely unspecified joint policy; nodes at depth 3 are fully specified.

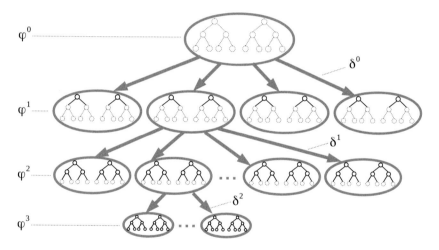

Fig. 4.4: An MAA* search tree.

MAA* performs standard A* search [Russell and Norvig, 2009]: it maintains an open list of partial joint policies φ_t and their heuristic values $\widehat{V}(\varphi_t)$. On every iteration MAA* selects the node φ_t with the highest heuristic value and 'expands' it, which means that it generates all child nodes—this involves generating and heuristically evaluating all $\varphi_{t+1} = \langle \varphi_t \circ \delta_t \rangle$ and placing them in the open list—and that it removes φ_t from the open list. When using an *admissible heuristic*—a guaranteed overestimation—the heuristic values $\widehat{V}(\varphi_{t+1})$ of the newly expanded policies are an upper bound to the true values, and any lower bound $\underline{v}$ that has been found (i.e., by expanding a full policy) can be used to prune the open list. The search ends when the list becomes empty, at which point an optimal fully specified joint policy has been found.

In order to compute a node's heuristic value $\widehat{V}(\varphi_t)$, MAA* takes $V^{0...t-1}(\varphi_t)$, the actual expected reward over the first t stages that are specified, and adds $\widehat{V}^{t...h-1}(\varphi_t)$, a heuristic value for the remaining $h-t$ stages. A typical way to specify $\widehat{V}^{t...h-1}(\varphi_t)$ is to use the value function of the *underlying MDP* [Szer et al., 2005]. That is, we can pretend that the Dec-POMDP is an MDP (by ignoring the observations and the decentralization requirement) and compute the (nonstationary) value function $V_{MDP}(s_t)$. Then we can use those values to specify

$$\widehat{V}^{t...h-1}(\varphi_t) = \sum_{s_t} \Pr(s_t | \varphi_t, b_0) V_{MDP}(s_t),$$

where the probability can be computed as the marginal of (3.4.4). As a result of its similarity to an approach to approximate POMDPs [Littman et al., 1995], this approach is called the Q_{MDP} heuristic. Intuitively, it should lead to an admissible heuristic because the MDP value function assumes that the state is observable. In a similar fashion, it is also possible to use the value function of the *underlying POMDP* [Szer et al., 2005, Roth et al., 2005a], referred to as Q_{POMDP}, which will also result in a tighter overestimation, since it makes fewer simplifying assumptions: it only drops the decentralization requirement, which effectively means that it assumes that the agents can communicate instantaneously (cf. Section 2.4.3). Finally, it is also possible to compute a heuristic, dubbed Q_{BG}, which corresponds to assuming that the agents can communicate with a one-step delay (see Section 8.3.2.1). For more details about these heuristics and empirical comparisons, we refer the reader to Oliehoek [2010]. Regardless of how $\widehat{V}^{t...h-1}(\varphi_t)$ is specified, when it is admissible—a guaranteed overestimation—so is $\widehat{V}(\varphi_t)$.

Even though MAA* improves over the brute-force search, its scalability is limited by the fact that the number of δ_t grows doubly exponential with t, which means that the number of children of a node grows doubly exponential in its depth. In order to mitigate the problem, it is possible to apply lossless clustering of the histories of the agents [Oliehoek et al., 2009], or to try and avoid the expansion of all child nodes by incrementally expanding nodes only when needed [Spaan et al., 2011]. These techniques can significantly improve the scalability of MAA* [Oliehoek et al., 2013a].

4.3 Converting to a Non-observable MDP

As discussed, the MAA* method stood at the basis of a number of advances in scalability of optimal methods. However, it also offers a new perspective on the planning process itself, which in turn has led to even greater improvements. We treat this perspective here.

4.3.1 The Plan-Time MDP and Optimal Value Function

In particular, it is possible to interpret the search-tree of MAA* as a special type of MDP, a *plan-time MDP*. That is, each node in Figure 4.4 (i.e., each past joint policy φ_t) can be interpreted to be a state and each edge (i.e., each joint decision rule δ_t) then corresponds to an action. In this plan-time MDP, the transitions are deterministic, and the rewards $\check{R}(\varphi_t, \delta_t)$ are the expected reward for stage t:

$$\check{R}(\varphi_t, \delta_t) = \mathbf{E}\left[R(s_t, a_t) \mid b_0, \varphi_t\right]$$
$$= \sum_{s_t}\sum_{\bar{o}_t}\Pr(s_t, \bar{o}_t | b_0, \varphi_t)R(s_t, \delta_t(\bar{o}_t)).$$

Here, the probability is given by a deterministic variant of (3.4.4).

The transitions are deterministic: the next past joint policy is determined completely by φ_t and δ_t, such that we can define the transitions as:

$$\check{T}(\varphi_{t+1}|\varphi_t, \delta_t) = \begin{cases} 1 & \text{if } \varphi_{t+1} = \langle \varphi_t \circ \delta_t \rangle, \\ 0 & \text{otherwise.} \end{cases}$$

Given that we have just defined an MDP, we can write down its optimal value function:

$$V_t^*(\varphi_t) = \max_{\delta_t} Q_t^*(\varphi_t, \delta_t) \tag{4.3.1}$$

where Q^* is defined as

$$Q_t^*(\varphi_t, \delta_t) = \begin{cases} \check{R}(\varphi_t, \delta_t) & \text{for the last stage } t = h-1, \\ \check{R}(\varphi_t, \delta_t) + V_{t+1}^*(\langle \varphi_t \circ \delta_t \rangle) & \text{otherwise.} \end{cases} \tag{4.3.2}$$

This means that, via the notion of plan-time MDP, we have been able to write down an optimal value function for the Dec-POMDP. It is informative to contrast the formulation of an *optimal* value function here to that of the value function *of a particular policy* as given by (3.4.2). Where the latter only depended on the history of observations, the optimal value function depends on the entire past joint policy. This means that, even though this optimal formulation admits a dynamic programming algorithm, it is not helpful, as this (roughly speaking) boils down to brute-force search through all joint policies [Oliehoek, 2010].

4.3.2 Plan-Time Sufficient Statistics

The problem in using the optimal value function defined by (4.3.1) is that it is too big: the number of past joint policies is too large to be able to compute it for most problems. However, it turns out that it is possible to replace the dependence on the past joint policy by a so-called *plan-time sufficient statistic:* a distribution over histories and states [Oliehoek et al., 2013a, Dibangoye et al., 2013]. This is useful, since many past joint policies can potentially map to the same statistic, as indicated in Figure 4.5.

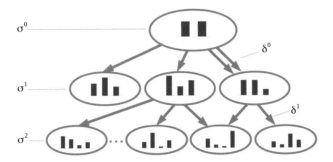

Fig. 4.5: A hypothetical MAA* search tree based on plan-time sufficient statistics. Two joint decision rules from the root node can map to the same σ_1, and two δ_1 (from different σ_1) can lead to the same σ_2.

Definition 22 (Sufficient Statistic for Deterministic Past Joint Policies). The sufficient statistic for a tuple (b_0, φ_t), with φ_t deterministic, is the distribution over joint observation histories and states: $\sigma_t(s_t, \bar{o}_t) \triangleq \Pr(s_t, \bar{o}_t | b_0, \varphi_t)$.

Such a statistic is sufficient to predict the immediate reward,

$$\check{R}(\sigma_t, \delta_t) = \sum_{s_t} \sum_{\bar{o}_t} \sigma_t(s_t, \bar{o}_t) R(s_t, \delta_t(\bar{o}_t)),$$

as well as the next statistic (a function of σ_t and δ_t). Let $\bar{o}_{t+1} = (\bar{o}_t, o_{t+1})$; then the updated statistic is given by

$$\sigma_{t+1}(s_{t+1}, \bar{o}_{t+1}) = U_{ss}(\sigma_t, \delta_t) = \sum_{s_t} \Pr(s_{t+1}, o_{t+1} | s_t, \delta_t(\bar{o}_t)) \sigma_t(s_t, \bar{o}_t). \quad (4.3.3)$$

This means that we can define the optimal value function for a Dec-POMDP as

$$V_t^*(\sigma_t) = \max_{\delta_t} Q_t^*(\sigma_t, \delta_t), \quad (4.3.4)$$

where

$$Q_t^*(\sigma_t, \delta_t) = \begin{cases} \check{R}(\sigma_t, \delta_t) & \text{for the last stage } t = h - 1, \\ \check{R}(\sigma_t, \delta_t) + V_{t+1}^*(U_{ss}(\sigma_t, \delta_t)) & \text{otherwise.} \end{cases} \quad (4.3.5)$$

Since potentially many φ_t map to the same statistic σ_t, the above formulation can enable a more compact representation of the optimal value function. Moreover, it turns out that this value function satisfies the same property as POMDP value functions:

Theorem 2 (PWLC of Optimal Value Function). *The optimal value function given by (4.3.4) is piecewise linear and convex (PWLC).*

Proof. For the proof see Oliehoek and Amato [2014].

4.3.3 An NOMDP Formulation

The PWLC property of the optimal value function seems to imply that we are actually dealing with a kind of POMDP. This intuition is correct [Nayyar et al., 2011, Dibangoye et al., 2013, MacDermed and Isbell, 2013]. In particular, it is possible make a reduction to a special type of POMDP: a non-observable MDP (a POMDP with just one 'NULL' observation).

Definition 23 (Plan-Time NOMDP). The *plan-time NOMDP* $\mathcal{M}_{PT}$ for a Dec-POMDP $\mathcal{M}_{DecP}$ is a tuple $\mathcal{M}_{PT}(\mathcal{M}_{DecP}) = \langle \check{\mathbb{S}}, \check{\mathbb{A}}, \check{T}, \check{R}, \check{\mathbb{O}}, \check{O}, \check{h}, \check{b}_0 \rangle$, where:

- $\check{\mathbb{S}}$ is the set of augmented states, each $\check{s}_t = \langle s_t, \bar{o}_t \rangle$.
- $\check{\mathbb{A}}$ is the set of actions, each $\check{a}_t$ corresponds to a joint decision rule δ_t in the Dec-POMDP.
- $\check{T}$ is the transition function:

$$\check{T}(\langle s_{t+1}, \bar{o}_{t+1} \rangle \,|\, \langle s_t, \bar{o}_t \rangle, \delta_t) = \begin{cases} \Pr(s_{t+1}, o_{t+1} | s_t, \delta_t(\bar{o}_t)) & \text{if } \bar{o}_{t+1} = (\bar{o}_t, o_{t+1}), \\ 0 & \text{otherwise.} \end{cases}$$

- $\check{R}$ is the reward function: $\check{R}(\langle s_t, \bar{o}_t \rangle, \delta_t) = R(s_t, \delta_t(\bar{o}_t))$.
- $\check{\mathbb{O}} = \{NULL\}$ is the observation set which only contains the *NULL* observation.
- $\check{O}$ is the observation function that specifies that *NULL* is received with probability 1 (irrespective of the state and action).
- The horizon is just the horizon of $\mathcal{M}_{DecP}$: $\check{h} = h$.
- $\check{b}_0$ is the initial state distribution. Since there is only one $\bar{o}_0$ (i.e., the empty joint observation history), it can directly be specified as

$$\forall s_0 \qquad \check{b}_0(\langle s_0, \bar{o}_0 \rangle) = b_0(s_0).$$

Since an NOMDP is a special case of a POMDP, all POMDP theory and solution methods apply. In particular, it should be clear that the belief in this plan-time NOMDP corresponds exactly to the plan-time sufficient statistic from Definition 22. Moreover, it can be easily shown that the optimal value function for this plan-time NOMDP is identical to the formulation equations (4.3.4) and (4.3.5).

4.4 Other Finite-Horizon Methods

Here we briefly describe two more methods for finite-horizon Dec-POMDPs. The first, point-based dynamic programming, is an extension of dynamic programming that tries to avoid work by only considering reachable multiagent beliefs. The second directly tries to transform the Dec-POMDP problem to a mathematical programming formulation. While these methods have been less effective on benchmark problems than the heuristic search and conversion to NOMDP methods discussed above, they present an insight into the problem and a basis for extensions.

4.4.1 Point-Based DP

The main problem in the scalability of exact dynamic programming, is that the set of maintained subtree policies grows very quickly. DP only removes q_i^τ that are not maximizing at *any point* in the multiagent belief space. *Point-based DP (PBDP)* [Szer and Charpillet, 2006] proposes improving pruning of the set $\mathbb{Q}_{e,i}^\tau$ by considering only a subset of *reachable* belief points $\mathbb{B}_i \subset \triangle(\mathbb{S} \times \mathbb{Q}_{-i}^\tau)$. Only those q_i^τ that maximize the value at some $b_i \in \mathbb{B}_i$ are kept.

In order to define reachable beliefs, we consider mappings Γ_j from observation histories to subtree policies: $\Gamma_{j,t} : \bar{\mathbb{O}}_{j,t} \to \mathbb{Q}_j^\tau$. Let $\Gamma_{-i,t} = \langle \Gamma_{j,t} \rangle_{j \neq i}$ be a mapping induced by the individual Γ_j. Now we can define the multiagent belief point induced by such a $\Gamma_{-i,t}$ and distribution $\Pr(s_t, \bar{o}_{-i,t} | b_0, \varphi_t)$ as follows:

$$b_{i,t}(s_t, q_{-i}^\tau) = \sum_{\bar{o}_{-i,t} \text{ s.t. } \Gamma_{-i,t}(\bar{o}_{-i,t}) = q_{-i}^\tau} \Pr(s_t, \bar{o}_{-i,t} | b_0, \varphi_t).$$

Definition 24 (Deterministically Reachable Multiagent Belief). A multiagent belief $b_{i,t}$ is reachable if there exists a probability distribution $\Pr(s_t, \bar{o}_{-i,t} | b_0, \varphi_t)$ (for any deterministic φ_t) and an induced mapping $\Gamma_{-i,t} = \langle \Gamma_{j,t} \rangle_{j \neq i}$ that result in $b_{i,t}$.

While PBDP has the potential to prune more and thus maintain smaller sets of subtree policies, its scalability remains limited since generating all reachable multiagent beliefs requires enumerating all the past joint policies φ_t. In order to overcome this problem, Szer and Charpillet [2006] also propose an approximate version that samples multiagent belief points. While we do not discuss this latter variant in detail, it can be interpreted as the starting point for an empirically very successful method, memory-bounded dynamic programming, which will be treated in Section 5.2.2.

4.4.2 Optimization

Another approach to solving Dec-POMDPs is to directly apply exact optimization methods. This is the approach taken by Aras and Dutech [2010], who proposed a mixed integer linear programming formulation for the optimal solution of finite-horizon Dec-POMDPs, based on representing the set of possible policies for each agent in *sequence form* [Koller and Pfeffer, 1997]. In this representation, a policy for an agent i is represented as a subset of the set of *sequences* (roughly corresponding to action-observation histories) for the agent. As such the problem can be interpreted as a combinatorial optimization problem—find the best subset of sequences—and solved with a mixed integer linear program (MILP).

For the formulation of the MILP, we refer the reader to the original paper of Aras and Dutech [2010]. We point out that, while the performance of the MILP-based approach has not been competitive with the newer MAA* variants, the link to optimization methods is an important one and may inspire future insights.

Chapter 5
Approximate and Heuristic Finite-Horizon Planning Methods

The previous chapter discussed methods for exactly solving finite-horizon Dec-POMDPs: i.e., methods that guarantee finding the optimal solution. While there have been quite a few insights leading to better scalability, finding an optimal solution remains very challenging and is not possible for many larger problems. In an effort to scale to these larger problems, researchers have considered methods that sacrifice optimality in favor of better scalability. Such methods come in two flavors: *approximation methods* and *heuristic methods*.

Approximation methods are not guaranteed to find the optimal solution, but have bounds on their solution quality. While such guarantees are very appealing, the complexity result by Rabinovich et al. [2003]—computing an ε-approximate joint policy is NEXP-complete; see Section 3.5—suggests that they may be either difficult to obtain, or will suffer from similar scalability problems as exact methods do. Looser bounds (e.g., probabilistic ones [Amato and Zilberstein, 2009]) are possible in some cases, but they suffer from the same trade-off between tightness and scalability.

Heuristic methods, in contrast, do not provide quality guarantees, but can produce high-quality results in practice. A difficulty with such methods is that it is unclear how close to optimal the solution is. This does not mean that such methods have no merit: they may produce at least some result where exact or approximation algorithms would not, and, even though we may not know how far from optimal they are, they can be objectively compared to one another on benchmark problems. Many communities in artificial intelligence and computer science have successfully used such benchmark-driven approaches. Moreover, we argue that the identification of a successful heuristic should not be seen as the end goal, but rather the start of an investigation into *why* that heuristic performs well: apparently there are some not-yet understood properties of the benchmarks that leads to the heuristic being successful.

Finally, let us point out that the terminology for non-exact methods is used loosely in the field. In order to avoid confusion with heuristic search methods (which, when used with an admissible heuristic—cf. Section 4.2.2—are optimal), heuristic methods are more often than not referred to as "approximate methods

© The Author(s) 2016
F.A. Oliehoek and C. Amato, *A Concise Introduction to Decentralized POMDPs*,
SpringerBriefs in Intelligent Systems, DOI 10.1007/978-3-319-28929-8_5

(without guarantees)". Consequently, approximation methods are often referred to as "bounded approximations" or "approximate methods with guarantees".

5.1 Approximation Methods

As indicated, approximation methods have not received much attention in Dec-POMDP literature. In fact, only one approximation method with quality guarantees has been proposed for finite-horizon Dec-POMDPs. This method is based on dynamic programming for Dec-POMDPs and will be treated first. However, as we will discuss next, the exact methods of the previous chapter can typically be transformed to approximation methods.

5.1.1 Bounded Dynamic Programming

Bounded Dynamic Programming (BDP) is a version of the dynamic programming algorithm from Section 4.1.2 that can compute a bounded approximation [Amato et al., 2007b]. It addresses the main bottleneck of DP, its memory requirement, by trading in memory for a bounded decrease in solution quality. The main idea is one that has also been employed in the approximation of POMDPs [Feng and Hansen, 2001, Varakantham et al., 2007a]: by allowing the pruning operator to more aggressively prune the sets of maintained policies, we lose some solution quality, but may potentially save a lot of space.

Instead of pruning only dominated policies as in the traditional dynamic programming approach, BDP performs a more aggressive ε-*pruning*. That is, instead of testing if $\upsilon \geq 0$ when pruning policies, it tests whether $\upsilon \geq -\varepsilon$ for some fixed error ε (thus, giving rise to the name ε-pruning). As a result, there may be some combination of policies that is only ε worse than the given policy q_i^τ, allowing the possible difference in value between the optimal set of policies and the ones retained to be bounded. Of course, truly suboptimal policies will also be pruned, but removing q_i^τ affects the value at that step by at most ε since pruning considers the value of policies for all distributions over states and other agent policies. Each time ε-pruning is used by any agent, policies that are close to the optimal may be lost, but the resulting value of the remaining policies can be bounded. Recall that in traditional dynamic programming, pruning at each step continues iteratively for each agent until no more policies can be removed. As a result, at time step τ, if ε-pruning is used n_τ times, the error is at most εn_τ. If ε-pruning is used at multiple time steps, the error will similarly increase additively: over h steps the error becomes at most $\varepsilon(n_1 + \ldots + n_h)$. Therefore, BDP can provide offline error bounds by interleaving ε-pruning with optimal pruning. For instance, pruning once per agent on each time step using ε-pruning, and using optimal pruning otherwise to ensure the total error is less then $n\varepsilon h$, where n is the number of agents and h is the horizon.

Online error bounds can also be calculated during BDP's execution to determine the actual values of υ that result in removing a policy. That is, if $\upsilon < 0$ during ε-pruning, the loss in value is at most $|\upsilon|$. The magnitude of the negative deltas can be summed to provide a more accurate estimate of the error that BDP has produced. Small epsilons are useful when implementing pruning in algorithms (due to numerical precision issues), but in practice large epsilons are often needed to significantly reduce the memory requirements of dynamic programming.

5.1.2 Early Stopping of Heuristic Search

The forward, heuristic search approach can also be transformed into an approximation method. In particular, remember that MAA* exhaustively searches the tree of past joint policies φ_t while maintaining optimistic heuristic values $\widehat{V}(\varphi_t)$. Since the method performs an A* search—it always selects the node to expand with the highest heuristic value—we know that if $\widehat{V}(\varphi_t^{selected})$ does not exceed the value $\underline{v}$ of the best found joint policy so far, it knows that is has identified an optimal solution. That is, MAA* stops when

$$\widehat{V}(\varphi_t^{selected}) \leq \underline{v}.$$

This immediately suggests a trivial way to turn MAA* into an algorithm that is guaranteed to find an ε-absolute error approximation: stop the algorithm when

$$\widehat{V}(\varphi_t^{selected}) \leq \underline{v} + \varepsilon.$$

Alternatively, MAA* can be used as an *anytime* algorithm: every time the algorithm finds a new best joint policy this can be reported. In addition, the value of $\widehat{V}(\varphi_t^{selected})$ can be queried at any time during the search, giving the current tightest upper bound, which allows us to compute the worst-case absolute error on the last reported joint policy.

5.1.3 Application of POMDP Approximation Algorithms

The conversion to a non-observable MDP (NOMDP), described in Section 4.3, immediately suggests an entire range of approximation methods. Even though there are not many methods directed at NOMDPs in particular, an NOMDP is a special case of a POMDP. As such, it is possible to apply any POMDP approximation method to the NOMDP formulation of a Dec-POMDP.

For instance, Dibangoye et al. [2013] extend heuristic search value iteration (HSVI) to apply to the NOMDP formulation. HSVI, introduced by Smith and Simmons [2004], performs a search over reachable beliefs given the initial belief, and maintains lower and upper bounds to direct the search. As such, in the limit, the lower and upper bound will converge, but the method can stop as soon as a desired precision ε is reached. In order to improve the efficiency of the method Dibangoye et al. [2013] introduce a number of modifications.

Because the state and action space of the NOMDP grows with the horizon, the authors compress these spaces by determining equivalent histories using methods such as the one discussed in Section 4.2.2. These compression techniques reduce the dimensionality of the sufficient statistic in the NOMDP. Using an extension of HSVI in this compressed space then permits significantly larger problems to be solved.

5.2 Heuristic Methods

Given the complexity of even finding an approximation for a Dec-POMDP, many proposed methods sacrifice guarantees on the performance in favor of better scalability. Again, many of these methods are inspired by the exact approaches (DP and MAA*) to solving Dec-POMDPs, but in addition there are a few other methods, of which we will describe *Joint Equilibrium-based Search for Policies* as well as cross-entropy optimization.

5.2.1 Alternating Maximization

Joint Equilibrium-based Search for Policies (JESP) is a method introduced by Nair et al. [2003c] that is guaranteed to find a *locally* optimal joint policy. More specifically, it is guaranteed to find a *Nash equilibrium*: a tuple of policies such that for each agent i its policy π_i is a best response for the policies employed by the other agents π_{-i}. Such a Nash equilibrium is a fixed point under best-response computation and is referred to as *person-by-person optimal* in control theory [Mahajan and Mannan, 2014].

JESP relies on a process called *alternating maximization*, which is a procedure that computes a policy π_i for an agent i that maximizes the joint reward, while keeping the policies of the other agents fixed. Next, another agent is chosen to maximize the joint reward by finding its best response. This process is repeated until the joint policy converges to a Nash equilibrium, which is a local optimum. This process is also referred to as *hill-climbing* or *coordinate ascent*. Note that the local optimum reached in this way can be arbitrarily bad. For instance, if, in the Dec-Tiger problem, agent 1 opens the left (a_{OL}) door right away, the best response for agent 2 is to also select a_{OL}. To reduce the impact of such bad local optima, JESP can use random restarts.

In order to compute a best response for the selected agent, one can exhaustively loop through the policies available to agent i and use (3.4.2) to compute the resulting value. However, a more efficient approach is to compute the best-response policy for a selected agent i via dynamic programming. In essence, fixing π_{-i} allows for a reformulation of the problem as an augmented POMDP. In this augmented POMDP, a state $\check{s} = \langle s, \bar{o}_{-i} \rangle$ consists of a nominal state s and the observation histories of the other agents $\bar{o}_{-i}$. Given the fixed deterministic policies of other agents π_{-i}, such an augmented state $\check{s}$ is Markovian and all transition and observation probabilities can be derived from π_{-i} and the transition and observation model of the original Dec-POMDP. For instance, the transition probabilities are given by

$$
\check{T}_{\pi_{-i}}(\check{s}' = \langle s_{t+1}, \bar{o}_{-i,t+1} \rangle | \check{s} = \langle s, \bar{o}_{-i,t} \rangle, a_{i,t}) =
$$
$$
T(s_{t+1}|s_t, \pi_{-i}(\bar{o}_{-i,t}), a_{i,t}) \sum_{o_{i,t+1}} O(o_{i,t+1}, o_{-i,t+1} | \pi_{-i}(\bar{o}_{-i,t}), a_{i,t}, s_{t+1})
$$

if $\bar{o}_{-i,t+1} = (\bar{o}_{-i,t}, o_{-i,t+1})$ and 0 otherwise. The observation probabilities of the augmented model are given by

$$
\check{O}(o_{i,t+1}|a_{i,t}, \check{s}' = \langle s_{t+1}, \bar{o}_{-i,t+1} \rangle) = O(o_{i,t+1}, o_{-i,t+1} | \pi_{-i}(\bar{o}_{-i,t}), a_{i,t}, s_{t+1}),
$$

where $o_{-i,t+1}$ is specified by $\bar{o}_{-i,t+1}$.

There is an interesting relation between the belief $b_i(\langle s, \bar{o}_{-i,t} \rangle)$ used by JESP, and a multiagent belief $b_i(\langle s, q_{-i}^{h-t} \rangle)$. Particularly, given a *full-length* (i.e., specified for all stages) fixed policy of other agents π_{-i}, it is easy to show that a 'JESP belief' induces a multiagent belief. Interestingly, it can be shown that (even when the full-length π_{-i} are not given) two histories that lead to the same JESP beliefs will lead to the same best-response policies for agent i [Oliehoek et al., 2013a]. This property can be used to cluster 'probabilistically equivalent' observation histories together (the 'lossless clustering' mentioned in Section 4.2.2), potentially allowing for drastic decreases in the search space.

JESP exploits the knowledge of the initial belief b_0 by only considering reachable beliefs $b_i(\check{s})$ in the solution of the POMDP. However, in some cases the initial belief might not be available. As demonstrated by Varakantham et al. [2006], JESP can be extended to plan for the entire space of initial beliefs, overcoming this problem.

5.2.2 Memory-Bounded Dynamic Programming

As discussed in Section 4.4.1, the major limitation of exact dynamic programming is the amount of time and memory required to generate and store all the nondominated subtree policies. Even though pruning is applied, this typically cannot prevent the sets of nondominated subtree policies from growing very rapidly. The Point-based DP technique that Section 4.4.1 introduces tries to overcome this problem by focusing only on *reachable multiagent beliefs,* and an approximate variation tries to bring further leverage by just sampling those reachable beliefs. In this section, we discuss a method called *Memory-bounded DP (MBDP)*, introduced by Seuken and Zilberstein [2007a] , which takes this idea even further.[1]

In particular, MBDP makes two approximations. First, for each agent it maintains sets of subtree policies $\mathbb{Q}_i^k$ of a fixed size: a parameter called *MaxTrees*. Second, in order to generate those *MaxTrees* subtree policies per agent, it does not sample reachable multiagent beliefs (which involves sampling past joint policies), but rather relies on a small number of heuristic joint policies and uses the sample *joint beliefs* (cf. Section 2.4.3) at which candidate subtree policies are evaluated. As such, MBDP integrates top-down heuristics with bottom-up dynamic programming.

In a bit more detail, MBDP is conducted in an iterative fashion like traditional dynamic programming. For example, in a problem of horizon h, the algorithm starts with performing an exhaustive backup to construct the sets $\{\mathbb{Q}_i^2\}$ of two-stage-to-go subtree policies. In order to reduce the size of these sets to *MaxTrees*, MBDP uses heuristic policies for stages $0,\ldots,h-2$ to sample joint-action observation histories $\bar{\theta}_{h-2}$ and associated joint beliefs b_{h-2}. At each joint belief the algorithm selects the joint subtree policy that maximizes the value:

$$q^2 = \arg\max_{q^2 \in \mathbb{Q}^2} V(b_{h-2}, q^2). \tag{5.2.1}$$

That is, MBDP pretends that the resulting joint beliefs are revealed to the agents and it retains only the trees that have the highest value at these joint belief. While during execution the belief state will not truly be revealed to the agents, the hope is that the *individual* subtree policies that are specified by these joint subtree policies are good policies in large parts of the multiagent belief space. Because it can happen that multiple maximizing joint subtree policies specify the same individual subtree for an agent, the algorithm continues sampling new joint beliefs b_{h-2} until it has found *MaxTrees* subtrees for each agent. At this point, MBDP will again perform an exhaustive backup and start with the selection of *MaxTrees* three-stage-to-go subtree policies for each agent.

The big advantage that MBDP offers is that, because the size of maintained subtrees does not grow, the size of the candidate sets $\mathbb{Q}_i^\tau$ formed by exhaustive backup is $O(|\mathbb{A}_\dagger| MaxTrees^{|\mathbb{O}_\dagger|})$, where $|\mathbb{A}_\dagger|$ and $|\mathbb{O}_\dagger|$ denote the size of the largest individual action and observation set. This does not depend on the horizon and as such MBDP scales linearly with respect to the horizon, enabling to solution of problems up to thousands of time steps.

While the complexity of MBDP becomes linear in the horizon, in order to perform the maximization in (5.2.1), MBDP loops over the

$$|\mathbb{Q}^\tau| = O(|\mathbb{A}_\dagger|^n MaxTrees^{n|\mathbb{O}_\dagger|})$$

joint subtree policies for each of the sampled joint belief points. To reduce the burden of this complexity, many papers have proposed new methods for performing

---

[1] For a deeper treatment of the relation between MBDP and PBDP, we refer to the description by Oliehoek [2012].

this so-called *point-based backup operation* [Seuken and Zilberstein, 2007b, Carlin and Zilberstein, 2008, Boularias and Chaib-draa, 2008, Dibangoye et al., 2009, Amato et al., 2009, Wu et al., 2010a].[2] Also, this backup corresponds to solving a one-shot constraint optimization problem, or collaborative Bayesian game, for each joint action [Kumar and Zilberstein, 2010a, Oliehoek et al., 2010].

5.2.3 Approximate Heuristic-Search Methods[3]

The fact that MBDP's point-based backup can be interpreted as a *collaborative Bayesian game (CBGs)* is no coincidence: such CBGs capture a fundamental aspect of the coordination process under uncertainty. In fact, one of the first approximate methods developed by Emery-Montemerlo et al. uses CBGs as a central component [Emery-Montemerlo et al., 2004, 2005]: it represents a Dec-POMDP as a series of CBGs, one for each stage t. It turns out that this method can be interpreted as a special case of the heuristic search approach to Dec-POMDPs (i.e., MAA*), providing a unifying perspective on the methods by Emery-Montemerlo et al. [2004] and Szer et al. [2005]. Here we treat this perspective, starting with a formalization of the CBG for a stage of the Dec-POMDP.

Dec-POMDPs as Series of Bayesian Games   A *Bayesian game (BG)* is an extension of a strategic form game in which the agents have private information [Osborne and Rubinstein, 1994]. A *collaborative* BG is a BG with identical payoffs. In a Dec-POMDP, the crucial difficulty in making a decision at some stage t is that the agents lack a common signal on which to condition their actions; they must instead base their actions on their individual histories. Given the initial state distribution b_0 and past joint policy φ_t, this situation can be modeled as a CBG:

Definition 25. A *collaborative Bayesian game* $B(\mathcal{M}_{DecP}, b_0, \varphi_t)$ for stage t of a Dec-POMDP $\mathcal{M}_{DecP}$ induced by b_0, φ_t is a tuple $\langle \mathbb{D}, \mathbb{A}, \bar{\Theta}_t, \Pr(\cdot), \widehat{Q} \rangle$ consisting of

- the set of agents $\mathbb{D}$,
- the set of their joint actions $\mathbb{A}$,
- the set of their joint AOHs $\bar{\Theta}_t$ (referred to as joint 'types' in Bayesian-game terminology),
- a probability distribution over them $\Pr(\bar{\theta}_t | b_0, \varphi_t)$, and
- a heuristic payoff function $\widehat{Q}(\bar{\theta}_t, a)$.

Since our discussion will be restricted to CBGs in the context of a single Dec-POMDP, we will simply write $B(b_0, \varphi_t)$ for such a CBG.

---

[2] This name indicates the similarity with the point-based backup in single-agent POMDPs.

[3] This section title nicely illustrates the difficulty with the terminology for approximate and heuristic methods: This section covers heuristic methods (i.e., without guarantees) that are based on heuristic search. In order to avoid the phrase 'heuristic heuristic-search methods', we will refer to these as 'approximate heuristic-search methods'.

In the CBG, agents use policies[4] that map from their individual AOHs to actions. That is, a policy of an agent i for a CBG corresponds to a decision rule $\delta_{i,t}$ for the Dec-POMDP. The solution of the CBG is the joint decision rule δ_t that maximizes the expected payoff with respect to $\widehat{Q}$:

$$\hat{\delta}_t = \underset{\delta_t}{\arg\max} \sum_{\bar{\theta}_t \in \bar{\Theta}_t} \Pr(\bar{\theta}_t | b_0, \varphi_t) \widehat{Q}(\bar{\theta}_t, \delta_t(\bar{\theta}_t)). \qquad (5.2.2)$$

Here $\delta_t(\bar{\theta}_t)$ is shorthand for the joint action resulting from individual application of the decision rules: $\delta_t(\bar{\theta}_t) \triangleq \langle \delta_{1,t}(\bar{\theta}_{1,t}), \ldots, \delta_{n,t}(\bar{\theta}_{n,t}) \rangle$. The probability is given as the marginal of (3.4.4). If φ_t is deterministic, the probability of $\bar{\theta}_t = \langle \bar{a}_t, \bar{o}_t \rangle$ is nonzero for exactly one $\bar{a}_t$, which means that attention can be restricted to OHs and decision rules that map from OHs to actions.

This perspective of a stage of a Dec-POMDP immediately suggests the following solution method: first construct a CBG for stage $t = 0$, solve it to find $\hat{\delta}_0$, set $\varphi_1 = (\hat{\delta}_0)$ and use it to construct a CBG $B(b_0, \varphi_1)$ for stage $t = 1$, etc. Once we have solved a CBG for every stage $t = 0, 1, \ldots, h-1$, we have found an approximate solution $\hat{\pi} = (\hat{\delta}_0, \ldots, \hat{\delta}_{h-1})$. This process is referred to as *forward-sweep policy computation (FSPC)* and is illustrated in Figure 5.1a.

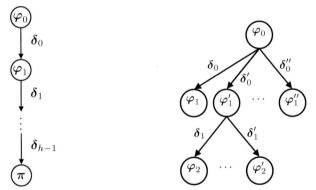

(a) Forward-sweep policy computation (FSPC). (b) (Generalized) MAA* performs backtracking.

Fig. 5.1: Forward approach to Dec-POMDPs.

A problem in FSPC is that (5.2.2) still maximizes over δ_t that map from histories to actions; the number of such δ_t is doubly exponential in t. There are two main approaches to gaining leverage. First, the maximization in (5.2.2) can be performed more efficiently: approximately via alternating maximization [Emery-Montemerlo et al., 2004], or exactly via heuristic search or other methods from constraint op-

---

[4] In game theory, these policies are typically referred to as 'strategies'. To avoid introducing more terms than necessary, we stick to policy here.

timization [Kumar and Zilberstein, 2010a, Oliehoek et al., 2010, 2012a]. Second, it is possible to reduce the number of histories under concern via pruning [Emery-Montemerlo et al., 2004], approximate clustering [Emery-Montemerlo et al., 2005] or lossless clustering [Oliehoek et al., 2009].

Heuristic Q-Value Functions The CBG for a stage is fully specified given b_0, φ_t and $\widehat{Q}$, but we have not yet addressed the matter of choosing $\widehat{Q}$. Essentially, this is quite similar to the choice of the heuristic $\widehat{V}$ in MAA* described in Section 4.2.2. The difference is that here we constrain the heuristic to be of the form:

$$\widehat{V}(\varphi_{t+1} = \langle \varphi_t \circ \delta_t \rangle) = \sum_{\bar{\theta}_t \in \bar{\Theta}_t} \Pr(\bar{\theta}_t | b_0, \varphi_t) \widehat{Q}(\bar{\theta}_t, \delta_t(\bar{\theta}_t)).$$

If, for the last stage, the heuristic specifies the immediate reward $\widehat{Q}(\bar{\theta}_t, a) = R(\bar{\theta}_t, a)$, it is easy to show that the decision rule $\hat{\delta}_{h-1}$ that maximizes (5.2.2) in fact maximizes the expected last-stage reward and thus is optimal (given b_0, φ_t). For other stages it is not practical to specify such an optimal heuristic of the form $\widehat{Q}(\bar{\theta}_t, a)$; this essentially corresponds to specifying an optimal value function, but there is no way to compute an optimal value function of such a simple form (cf. the discussion in Section 4.3).

However, note that FSPC via CBGs is not suboptimal per se: It is possible to compute a value function of the form $Q^\pi(\bar{\theta}_t, a)$ for any π. Doing this for a π^* yields Q^{π^*}, and when using the latter as the payoff functions for the CBGs, FSPC is exact [Oliehoek et al., 2008b].[5] The practical value of this insight is limited since it requires knowing an optimal policy to start with. In practice, researchers have used approximate value functions, such as the Q_{MDP}, Q_{POMDP} and Q_{BG} functions that were mentioned in Section 4.2.2. It is worth pointing out, however, that since FSPC does not give any guarantees, it is not restricted to using an 'admissible' heuristic: heuristics that occasionally underestimate the value but are overall more accurate can produce better results.

Generalized MAA* Even though Figure 5.1 shows a clear relation between FSPC and MAA*, it may not be directly obvious how they relate: the former solves CBGs, while the latter performs heuristic search. Generalized MAA* (GMAA*) unifies these two approaches by making explicit the 'Next' operator [Oliehoek et al., 2008b].

Algorithm 5.1 shows GMAA*. When the Select operator selects the highest ranked φ_t and when the expansion ('Next') operator expands all the children of a node, GMAA* simply is MAA*. Alternatively, the Next operator can construct a CBG $B(b_0, \varphi_t)$ for which all joint CBG policies δ_t are evaluated. These can then be used to construct a new set of partial policies $\Phi_{\text{Next}} = \{\langle \varphi_t \circ \delta_t \rangle\}$ and their heuristic values. This corresponds to MAA* reformulated to work on CBGs. It can be shown

---

[5] There is a subtle but important difference between $Q^{\pi^*}(\bar{\theta}_t, a)$ and the optimal value function from Section 4.3: the latter specifies the optimal value given *any* past joint policy φ_t while the former only specifies optimal value given that π^* is actually being followed. For a more thorough discussion of these differences we refer you to Oliehoek [2010].

Algorithm 5.1 (Generalized) MAA*

Initialize:
$\underline{v} \leftarrow -\infty$ {initialize max. lower bound}
$L \leftarrow \{\varphi_0 = ()\}$ {initialize open list with 'empty' root node}
repeat
    $\varphi_t \leftarrow \texttt{Select}(L)$
    $\Phi_{\texttt{Next}} \leftarrow \texttt{Next}(b_0, \varphi_t)$
    if $\Phi_{\texttt{Next}}$ contains full policies $\Pi_{\texttt{Next}} \subseteq \Phi_{\texttt{Next}}$ **then**
        $\pi' \leftarrow \arg\max_{\pi \in \Pi_{\texttt{Next}}} V(\pi)$
        if $V(\pi') > \underline{v}$ **then**
            $\underline{v} \leftarrow V(\pi')$ {found better lower bound}
            $\pi^\star \leftarrow \pi'$
            $L \leftarrow \{\varphi \in L \mid \widehat{V}(\varphi) > \underline{v}\}$ {prune L}
        end if
        $\Phi_{\texttt{Next}} \leftarrow \Phi_{\texttt{Next}} \setminus \Pi_{\texttt{Next}}$ {remove full policies}
    end if
    $L \leftarrow (L \setminus \varphi_t) \cup \{\varphi \in \Phi_{\texttt{Next}} \mid \widehat{V}(\varphi) > \underline{v}\}$ {remove processed/add new φ }
until L is empty

---

that when using a particular form of $\widehat{Q}$ ($\widehat{Q}$ needs to faithfully represent the expected immediate reward; the mentioned heuristics Q_{MDP}, Q_{POMDP} and Q_{BG} all satisfy this requirement), the approaches are identical [Oliehoek et al., 2008b]. GMAA* can also use a $\texttt{Next}$ operator that does not construct all new partial policies, but only the best-ranked one, $\Phi_{\texttt{Next}} = \{\langle \varphi_t \circ \delta_t^* \rangle\}$. As a result the open list L will never contain more than one partial policy, and behavior reduces to FSPC. A generalization called k-GMAA* constructs the k best-ranked partial policies, allowing us to trade off computation time for solution quality. Clustering of histories can also be applied in GMAA*, but only lossless clustering will preserve optimality.

5.2.4 Evolutionary Methods and Cross-Entropy Optimization

The fact that solving a Dec-POMDP can be approached as a combinatorial optimization problem, as discussed in Section 4.4.2, was also recognized by approaches based on cross-entropy optimization [Oliehoek et al., 2008a] and genetic algorithms [Eker and Akın, 2010]. Here we given an overview of *direct cross-entropy* (DICE) optimization for Dec-POMDPs based on the article by Oliehoek et al. [2008a].

The cross-entropy method can be used for optimization in cases where we want to find a—typically large—vector x from a hypothesis space $\mathscr{X}$ that maximizes some *performance function* $V : \mathscr{X} \to \mathbb{R}$. That is, when we are looking for

$$x^* = \arg\max_{x \in \mathscr{X}} V(x). \tag{5.2.3}$$

The CE method maintains a probability distribution f_ξ over the hypothesis space, parametrized by a vector ξ.

The core of the CE method is an iterative two-phase process:

1. Generate a set of samples $\mathbf{X}$ according to f_ξ.
2. Select the best N_b samples $\mathbf{X}_b$, and use those to update the parameter vector ξ.

DICE is a translation of this idea to finite-horizon Dec-POMDPs. In the Dec-POMDP case, the hypothesis space is the space of deterministic joint policies Π. In order to apply the CE method, it is required to define a distribution over this space, an evaluation function for sampled policies, and the manner by which the policy distribution is updated from the best samples.

Policy Distribution Let f_ξ denotes a probability distribution over pure joint policies, parametrized by ξ. It is the product of probability distributions over individual pure joint policies:

$$f_\xi(\pi) = \prod_{i=1}^{n} f_{\xi_i}(\pi_i). \qquad (5.2.4)$$

Here ξ_i is the vector of parameters for agent i, i.e., $\xi = \langle \xi_1,...,\xi_n \rangle$.

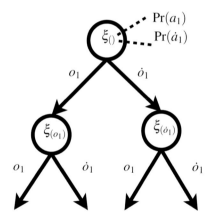

Fig. 5.2: A part of a stochastic policy for an agent of a fictitious Dec-POMDP.

The question is how to represent the probability distributions over individual pure policies. The simplest way is illustrated in Figure 5.2: maintaining a simple probability distribution over actions for each observation history. Other choices, such as maintaining distributions over actions for every AOH or controller node, are also possible.

Sampling and Evaluation Unlike other applications of the CE method (e.g., Mannor et al. 2003), in the setting of Dec-POMDPs there is no trivial way to sample trajectories given the joint policy distribution f_ξ and use that to update the distribu-

Algorithm 5.2 The DICE policy search algorithm

Input: CE parameters: I, N, N_b, α

1: $V_b \leftarrow -\infty$
2: initialize $\xi^{(0)}$ {typically uniform random}
3: **for** $i \leftarrow 0$ to I **do**
4:     $\mathbf{X} \leftarrow \emptyset$
5:     **for** $s \leftarrow 0$ to N **do**
6:         sample π from $f_{\xi^{(i)}}$
7:         $\mathbf{X} \leftarrow \mathbf{X} \cup \{\pi\}$
8:         $V(\pi) \leftarrow \text{Evaluate}(\pi)$ {exactly or approximately}
9:         **if** $V(\pi) > V_b$ **then**
10:           $V_b \leftarrow V(\pi)$
11:           $\pi_b \leftarrow \pi$
12:         **end if**
13:     **end for**
14:     $\mathbf{X}_b \leftarrow$ the set of N_b best joint policies $\pi \in \mathbf{X}$
15:     Compute $\xi^{(i+1)}$ {using (5.2.5) }
16:     $\xi^{(i+1)} \leftarrow \alpha \xi^{(i+1)} + (1 - \alpha)\xi^{(i)}$
17: **end for**
18: **return** π_b

---

tion. Rather DICE samples complete joint policies and uses those for the parameter update.

Selecting a random sample $\mathbf{X}$ of N joint policies π from the distribution f_ξ is straightforward. For all the observation histories $\bar{o}_{i,t}$ of an agent i an action can be sampled from action distribution $\xi_{\bar{o}_{i,t}}$. The result of this process is a deterministic policy for agent i. Repeating this procedure for each agent samples a deterministic joint policy. The evaluation of a joint policy can be done using (3.4.2). For larger problems—where policy evaluation is expensive—it is also possible to do approximate sample-based evaluation using only polynomially many samples [Oliehoek et al., 2008b].

Parameter Update   The final step is to update the distribution using the best joint policies sampled. Let $\pi \in \mathbf{X}_b$ be the set of best joint policies sampled from the previous distribution $f_{\xi^{(j)}}$. These will be used to find new parameters, $\xi^{(j+1)}$. Let $\mathbf{1}_{\{\pi_i(\bar{o}_{i,t})\}}(a_i)$ be an indicator function that indicates whether $\pi_i(\bar{o}_{i,t}) = a_i$. In the OH-based distribution the probability of agent i taking action $a_{i,t}$ after having observed $\bar{o}_{i,t}$ can be re-estimated as:

$$\xi^{(j+1)}_{\bar{o}_{i,t}}(a_i) = \frac{1}{|\mathbf{X}_b|} \sum_{\pi \in \mathbf{X}_b} \mathbf{1}_{\{\pi_i(\bar{o}_{i,t})\}}(a_i), \qquad (5.2.5)$$

where $|\mathbf{X}_b|$ normalizes the distribution. Note that the computed new parameter vector $\xi^{(j+1)}$ can be smoothed using a learning rate parameter α.

Summary   Algorithm 5.2 summarizes the DICE policy search method. To start, it needs I, the number of iterations, N, the number of samples taken at each iteration,

N_b, the number of samples used to update ξ, and α, the learning rate. The outer loop of lines 3–17 covers one iteration. The inner loop of lines 5–13 covers sampling and evaluating one joint policy. Lines 14–16 perform the parameter update. Because the CE method can get stuck in local optima, one typically performs a number of restarts. This algorithm has also been extended to solve the infinite-horizon problems discussed in the next chapter [Omidshafiei et al., 2016].

Chapter 6
Infinite-Horizon Dec-POMDPs

This chapter presents an overview of the theory and policy representations for infinite-horizon Dec-POMDPs. The optimality criteria are discussed and policy representations using finite-state controllers are considered. These controllers can be thought of as an extension of the policy trees used in the finite-horizon case to allow execution over an infinite number of steps. This chapter provides value functions for such controllers. Furthermore, the decidability and computational complexity of the problem of (ε-)optimally solving infinite-horizon Dec-POMDPs are discussed.

6.1 Optimality Criteria

When an infinite horizon is considered, the expected cumulative reward may be infinite. A finite sum (and a well-defined optimization problem) can be maintained by using discounting or average rewards. Both of these concepts are presented below, but the vast majority of research considers the discounted cumulative reward case.

6.1.1 Discounted Cumulative Reward

In the infinite horizon case, a finite sum can be guaranteed by considering the discounted expected cumulative reward

$$DECR = \mathbf{E}\left[\sum_{t=0}^{\infty} \gamma^t R(s_t, a_t)\right], \tag{6.1.1}$$

where $0 \leq \gamma < 1$ is the discount factor. That is, the DECR here is exactly the same as in the finite horizon (3.1.3), even though we explicitly replaced the summand with ∞ here. Where in the finite-horizon setting discounting might be applied if it makes sense from the application perspective to give more importance to earlier rewards,

© The Author(s) 2016
F.A. Oliehoek and C. Amato, *A Concise Introduction to Decentralized POMDPs*,
SpringerBriefs in Intelligent Systems, DOI 10.1007/978-3-319-28929-8_6

in the infinite-horizon discounting is (also) applied to make sure that the objective is bounded. In this way, discounting makes the values of different joint policies comparable even if they operate for an infinite amount of time.

6.1.2 Average Reward

A different way to overcome the unbounded sum that would result from regular expected cumulative rewards is given by the expected average reward criterion:

$$EAR = \mathbf{E}\left[\lim_{h\to\infty} \frac{1}{h} \sum_{t=0}^{h-1} R(s_t, a_t)\right], \qquad (6.1.2)$$

This criterion has the benefit that it does not require setting a discount factor. In contrast to DECR, earlier states do not weight more heavily and, in fact, any finite sequence of initially poor rewards will be disregarded by this criterion: only the limiting performance counts. Therefore, it is most appropriate for problems that are truly expected to run for what can be considered an infinite amount of time (as opposed to problems that must reach some goal or complete some set of tasks in an *unknown* amount of time). Theoretical analysis of the average reward criterion in the POMDP case is very involved and the complexity of the problem is the same as that of the discounted case (undecidable) [Puterman, 1994]. Few researchers have considered the average reward case, but it has been shown to be NP-complete in the case of independent transition and observation Dec-MDPs [Petrik and Zilberstein, 2007] and has been used in conjunction with the expectation maximization methods described by Pajarinen and Peltonen [2013]. Since the amount of work done on the average-reward case is limited, we will focus on the discounted cumulative reward criterion in this book.

6.2 Policy Representation

A tree-based representation of a policy requires the agent to perfectly remember the entire history in order to determine its next action, which in the case of an infinite-horizon problem would imply that an agent needs an infinite amount of memory. Clearly, this is not possible. As an alternative, we discuss approaches that consider finite sets of internal states $\mathbb{I}_i$ which represent an agent i's finite memory.

A natural representation for a policy in this form is given by *finite-state controllers (FSCs)*. FSCs can be used to represent policies for agents in an elegant way since an agent can be conceptualized as a device that receives observations and produces actions. FSCs operate in a manner that is very similar to policy trees in that there is a designated initial node, and following the action selection at that node, the controller transitions to the next node depending on the observation seen. This continues for the infinite steps of the problem. Nodes in the controller of agent i represent its internal states $\mathbb{I}_i$ and prescribe actions based on this finite memory.

For Dec-POMDPs, a set of controllers, one per agent, provides the joint policy. Finite-state controllers explicitly represent infinite-horizon policies, but can also be used (as a possibly more concise representation) for finite-horizon policies. They have been widely used in POMDPs[1] (e.g., see Kaelbling et al. 1998, Hansen 1998, Meuleau et al. 1999b, Poupart and Boutilier 2004, Poupart 2005, Toussaint et al. 2006, 2008, Grześ et al. 2013, Bai et al. 2014) as well as Dec-POMDPs (e.g., Bernstein et al. 2005, Szer and Charpillet 2005, Amato et al. 2007a, Bernstein et al. 2009, Kumar and Zilberstein 2010b, Pajarinen and Peltonen 2011a, Kumar et al. 2011, Pajarinen and Peltonen 2011b and Wu et al. 2013).

One thing to note is that compared to the finite-horizon setting treated in the previous chapters, introducing FSCs somewhat alters the multiagent decision problem that we are dealing with. The previous chapters assumed that agents' actions are based on histories, thereby in fact (implicitly) specifying the agents' belief update function. When resorting to FSCs this is no longer the case, and we will need to reason about both the action selection policies π_i as well as the belief update functions ι_i.

6.2.1 Finite-State Controllers: Moore and Mealy

Perhaps the easiest way to view a finite-state controller (FSC) is as an agent model, as treated in Section 2.4.4, where the number of internal states (or simply 'states') is finite. The typical notation employed for FSCs is different from the notation from Section 2.4.4—these differences are summarized in Table 6.1— but we hope that the parallel is clear and will reuse the agent model's notation for FSCs (which are an instantiation of such agent models).[2] To differentiate internal states of the FSC from states of the Dec-POMDP, we will refer to internal controller states as *nodes*.

We will now focus on two main types of FSCs, Moore and Mealy: Moore controllers associate actions with nodes and Mealy controllers associate actions with controller transitions (i.e., nodes and observations). The precise definition of the components of FSCs can be formulated in different ways (e.g., deterministic vs. stochastic transition functions, as will be further discussed in the remainder of this section).

Definition 26. A (deterministic) *Moore controller* for agent i can be defined as a tuple $m_i = \langle \mathbb{I}_i, I_{i,0}, \mathbb{A}_i, \mathbb{O}_i, \iota_i, \pi_i \rangle$ where

---

[1] In POMDPs, finite-state controllers have the added benefit (over value function representations) that the policy is explicitly represented, alleviating the need for belief updating during execution.

[2] In this section, we will not consider auxiliary observations, and thus omit $\mathbb{Z}_i$ from the definitions. Note that FSCs are not per se incompatible with auxiliary observations: they could be allowed by defining the Cartesian product $\mathbb{O}_i \times \mathbb{Z}_i$ as the input alphabet in Table 6.1.

model $m_i = \langle \mathbb{I}_i, I_i, \mathbb{A}_i, \mathbb{O}_i, \iota_i, \pi_i \rangle$	FSC $\langle \mathbb{N}, n_0, \Sigma, \Upsilon, \beta, \alpha \rangle$	mixed
$\mathbb{I}_i$ a set of internal states	S finite set of states	$\mathbb{Q}_i$
$I_{i,0}$ initial internal state	s_0 initial state	q_i
$\mathbb{O}_i$ the set of observations	Σ the finite input alphabet	$\mathbb{O}_i$
$\mathbb{A}_i$ the set of actions	Γ the finite output alphabet	$\mathbb{A}_i$
π_i the action selection policy	λ the output function	λ_i
ι_i the belief update function	δ the transition function	δ_i

Table 6.1: Mapping between model notation from Section 2.4.4 and (typical) finite-state machine notation and terminology, as well as a 'mixed' notation common in Dec-POMDP literature.

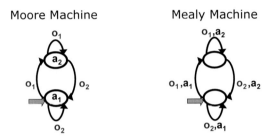

Fig. 6.1: Deterministic Moore and Mealy finite-state controllers.

- $\mathbb{I}_i$ is the finite set of nodes,
- $I_{i,0}$ is the initial node,
- $\mathbb{O}_i$ and $\mathbb{A}_i$ are the input and output alphabets,
- $\iota_i : \mathbb{I}_i \times \mathbb{O}_i \to \mathbb{I}_i$ is the (deterministic) transition function,[3]
- $\pi_i : \mathbb{I}_i \to \mathbb{A}_i$ is the (deterministic) output function.

In the Moore case, execution begins at the given initial node, $I_{i,0}$, the action associated with that node, $\pi_i(I_{i,0})$, is selected and then the controller transitions to a new node based on the observation seen by using the controller transition (i.e., belief update) function: $I_{i,1} = \iota_i(I_{i,0}, o_{i,1})$. These action selections and controller transitions can continue for the infinite steps of the problem.

Definition 27. A *Mealy controller* is a tuple $m_i = \langle \mathbb{I}_i, I_{i,0}, \mathbb{A}_i, \mathbb{O}_i, \iota_i, \pi_i, \pi_{i,0} \rangle$ where

- $\mathbb{I}_i, I_{i,0}, \mathbb{A}_i,$ and $\mathbb{O}_i$ are as in a Moore controller
- $\pi_{i,0} : \mathbb{I}_i \to \mathbb{A}_i$ is the output function for the first stage $t = 0$,
- $\pi_i : \mathbb{I}_i \times \mathbb{O}_i \to \mathbb{A}_i$ is the output function for all the remaining stages.

The difference in the Mealy case is that action selection now depends on the last observation seen as well as the current node. As such, a separate rule is needed for

---

[3] Note that the transition function (for both the Moore and Mealy case) can also depend on the action chosen, which makes the formulation even more general. All the algorithms in the next chapter can operate in either case.

the first stage (the action selection policy for this stage cannot depend on an observation since none have been seen yet). Examples of two-node Moore and Mealy controllers are shown in Figure 6.1.

Both Moore and Mealy models are equivalent in the sense that for a given controller of one type, there is a controller of the other type that generates the same outputs. However, it is known that Mealy controllers are more succinct than Moore controllers in terms of the number of nodes. Given a Moore controller m_i, one can find an equivalent Mealy controller m_i' with the same number of nodes by constraining the outputs produced at each transition from a common node to be the same. Conversely, given a (general) Mealy controller, the equivalent Moore controller has $|\mathbb{I}_i| \times |\mathbb{O}_i|$ nodes [Hopcroft and Ullman, 1979]. Of course, more parameters are needed for a Mealy controller ($2|\mathbb{I}_i| \times |\mathbb{O}_i|$ in the Mealy case, but only $|\mathbb{I}_i| + |\mathbb{I}_i| \times |\mathbb{O}_i|$ in the Moore case), but this added structure can be used by algorithms (e.g., limiting the possible actions considered based on the observation seen at the node). In general, both formulations can be useful in solution methods (as we will discuss in the next chapter).

6.2.2 An Example Solution for DEC-TIGER

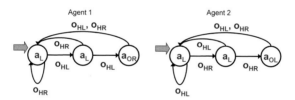

Fig. 6.2: Three node deterministic controllers for two agents in the DEC-TIGER problem.

An example of a set of Moore controllers for the (two agent) DEC-TIGER problem is given in Figure 6.2. This is the highest quality deterministic solution which uses at most three nodes for each agent. Here, agent 1 listens until it hears the tiger on the left twice in a row and then chooses to open the door on the right, while agent 2 listens until it hears the tiger on the right twice in a row and then opens the door on the left. After the door is opened, the agents transition back to the first node and begin this process again. The value of these controllers (using a discount factor of 0.9) is approximately -14.12, while the value of listening forever is -20.

6.2.3 Randomization

Even though switching to FSCs helps us overcome the problem of requiring an infinite amount of memory, deterministic controllers limit the value of policies represented as controllers. That is, with a limited amount of memory, stochastic controllers are able to produce higher-quality solutions with the same number of nodes compared to deterministic controllers [Singh et al., 1994, Bernstein et al., 2009]. This makes stochastic controllers useful for methods that seek to keep the number of nodes small, but stochastic controllers can always be converted to (larger) deterministic controllers [Bernstein et al., 2009].

Stochastic controllers are similar to those presented above except that probability distributions are used for transitions and output. That is, for a Moore controller, the transition function is $\iota_i : \mathbb{I}_i \times \mathbb{O}_i \to \triangle(\mathbb{I}_i)$ (where $\triangle(\mathbb{I}_i)$ is the set of probability distributions over $\mathbb{I}_i$) and the output function is $\pi_i : \mathbb{I}_i \to \triangle(\mathbb{A}_i)$. For Mealy controllers, these become $\iota_i : \mathbb{I}_i \times \mathbb{O}_i \to \triangle(\mathbb{I}_i)$ and $\pi_i : \mathbb{I}_i \times \mathbb{O}_i \to \triangle(\mathbb{A}_i)$.

6.2.4 Correlation Devices

Adding randomization may allow higher quality solutions to be represented more concisely in Dec-POMDPs. However, because policies need to be decentralized, we are not able to represent all joint probability distributions, but only factored ones. In other words, since agent policies depend only on local action and observation histories, even higher quality solutions could be achieved if we allow the action selection and node transitions to be correlated between agents.

Specifically, controllers have also been developed that make use of a shared source of randomness in the form of a *correlation device*. This allows a set of independent controllers to be correlated in order to produce higher values, without sharing any additional local information. For example, consider a situation in which all agents are able to observe the outcome of a coin that is being flipped on each step. The agents can then condition their action choices and controller transitions on these outcomes to produce correlated behavior. Specifically, the correlation device is a tuple $\langle \mathbb{C}, \psi \rangle$, where $\mathbb{C}$ is a set of correlation device states and $\psi : \mathbb{C} \to \triangle(\mathbb{C})$ is a stochastic transition function that we will represent as $\Pr(c'|c)$. At each step of the problem, the device transitions and each agent can observe its state. Then, for a Moore controller, the action selection probabilities can be defined as $\pi_i : \mathbb{I}_i \times \mathbb{C} \to \triangle(\mathbb{A}_i)$ while the node transition probabilities an be defined as $\iota_i : \mathbb{I}_i \times \mathbb{O}_i \times \mathbb{C} \to \triangle(\mathbb{I}_i)$. A correlation can be incorporated into a Mealy controller in a similar way. It is worth noting that correlation devices are similar to non-influenceable state factors, s_0, that were discussed in Section 2.4.2 . Both of their state values are uninfluenceable, but agents can condition their actions on them in order to correlate their solutions (and possibly improve their performance).

A correlation device is particularly useful for reducing miscoordination in stochastic controllers since the agents can choose the same (or appropriate) action based on the commonly known signal. For instance, consider a domain in which there is a large penalty when agents choose different actions, but a large reward for choosing the same actions. For sufficiently small controllers (e.g., one node for each agent), this type of policy is impossible without the correlation device. It has been shown that policies can be randomized and correlated to allow higher values to be attained in a range of domains [Bernstein et al., 2005, 2009, Amato et al., 2007a, 2010].

6.3 Value Functions for Joint Policies

When each agent uses a Moore controller, this results in a fully specified agent component m (cf. Definition 8), to which we will also refer as *joint controller* in the current context of FSCs. Such a joint controller induces a Markov reward process (which is a Markov chain with rewards, or, alternatively, an MDP without actions) and thus a value. In particular, the infinite-horizon discounted reward incurred when the initial state is s and the initial nodes for all of the controllers is given by I can be denoted by $V^m(I,s)$ and satisfies:

$$V^m(I,s) = \sum_a \pi(a|I) \left(R(s,a) + \gamma \sum_{s',o,I'} \Pr(s',o|s,a) \iota(I'|I,o) V^m(I',s') \right) \quad (6.3.1)$$

where

- $\pi(a|I) \triangleq \prod_i \pi_i(a_i|I_i)$, and
- $\iota(I'|I,o) \triangleq \prod_j \iota_i(I_i'|I_i,o_i)$.

The value of m at the initial distribution is $V^m(b_0) = \sum_{s_0} b_0(s_0) V^m(I_0,s_0)$, where I_0 is the set of initial nodes for the agents.

For a Mealy controller, the selected actions will depend on the last received observations. For a particular joint controller, m, when the initial state is s, the last joint observation was o, and the current node of m is I, the value is denoted by $V^m(I,o,s)$ and satisfies:

$$V^m(I,o,s) = \pi(a|I,o) \left(R(s,a) + \gamma \sum_{s',o,I'} \Pr(s',o|s,a) \iota(I'|I,o) V^m(I',o',s') \right) \quad (6.3.2)$$

where, now, $\pi(a|I,o) = \prod_i \pi_i(a_i|I_i,o_i)$. In this case, recall that the first node is assumed to be a Moore node (i.e., the action selection in the first stage is governed by $\pi_{i,0}$, which only depends on the node, not on observations),[4] so the value for the initial belief b_0 can be computed as $V^m(b_0) = \sum_{s_0} b_0(s_0) V_0^m(I_0,s_0)$, with

---

[4] Alternatively, the value can be represented as $V^m(b_0) = \sum_{s_t} b_0(s_0) V_0^m(I_0,o_0^* s_0))$, where o_0* is a dummy observation that is only received on the first step of the problem.

$$V_0^m(I_0,s_0) = \sum_a \pi_0(a|I) \left(R(s,a) + \gamma \sum_{s',o,I'} \Pr(s',o|s,a)\iota(I'|I,o)V^m(I',o',s') \right).$$

These recursive equations are similar to those used in the finite-horizon version of the problem, but evaluation continues for an infinite number of steps. Note that the policy is now stationary (depending on the controller node, but not time) and the value for each combination of nodes and states for a fixed policy can be found using a set of linear equations or iterative methods [Amato et al., 2010, Bernstein et al., 2009].

6.4 Undecidability, Alternative Goals and Their Complexity

For the infinite-horizon problem, both the number of steps of the problem and the possible size of the policy (i.e., the number of nodes in the controllers) are unbounded. That is, controllers of unbounded size may be needed to perfectly represent an optimal policy. As a result, solving an infinite-horizon Dec-POMDP optimally is undecidable. This follows directly from the fact that optimally solving infinite-horizon POMDPs is undecidable [Madani et al., 1999], since a Dec-POMDP is a generalization of a POMDP.

Similarly, the definition of multiagent beliefs, which is based on subtree policies (cf. Definition 19), is not appropriate in the potentially infinite space, but has been reformulated in the context of bounded policies. Specifically, from the perspective of agent i and given a known set of controllers for the other agents, $-i$, a probability distribution of the other agents being in nodes I_{-i} while the state of the system is s_t, can be represented as $\Pr(s_t, I_{-i})$. Like the multiagent belief in the finite-horizon case, these probabilities can be used at planning time to evaluate agent i's policies across the space of other agent policies and estimate the outcomes of other agents' controllers.

As an alternative, approximation methods (with guarantees) have been considered: due to discounting, a solution that is within any fixed ε of the optimal value can be found in a finite number of steps [Bernstein et al., 2009]. That is, we can choose t such that the maximum sum of rewards over the remaining stages $t+1, t+2, \ldots$ is bounded by ε:

$$\sum_{k=t+1}^{\infty} \gamma^k |R_{\max}| = \frac{\gamma^{t+1}|R_{\max}|}{1-\gamma} \leq \varepsilon,$$

where $|R_{max}|$ is the immediate reward with the largest magnitude. This ensures that any sum of rewards after time t will be smaller than ε (due to discounting). Therefore, t becomes an *effective* horizon in that an optimal solution for the horizon t problem ensures an ε-optimal solution for the infinite-horizon problem. This procedure is also NEXP-complete because an optimal policy is found for the effective horizon. Theoretically, finite-horizon methods from the previous chapter could also be used to produce ε-optimal solutions, but the effective horizon is often too large for current optimal finite-horizon methods to reach. As a more scalable solution, heuristic methods have been developed which seek to produce the highest quality solution for a given controller size. These methods are discussed in more detail in the following chapter.

Chapter 7
Infinite-Horizon Planning Methods: Discounted Cumulative Reward

As discussed in Chapter 6, infinite-horizon planning methods focus on generating ε-optimal solutions or solutions with a fixed controller size. We first discuss policy iteration, which is an extension of dynamic programming for Dec-POMDPs to the infinite-horizon case and can produce ε-optimal solutions. We then describe some of the heuristic algorithms that generate fixed-size controllers. We describe these algorithms in terms of Moore controllers, but they can all use Mealy controllers (see Section 6.2) with minor extensions.

7.1 Policy Iteration

Policy iteration (PI) for Dec-POMDPs [Bernstein et al., 2009] is similar to the finite-horizon dynamic programming algorithm (Algorithm 4.1), but finite-state controllers are used as policy representations (like the policy iteration approach for POMDPs [Hansen, 1998]). In addition, there are two other main differences with the finite-horizon DP:

1. Instead of using a set of separately represented policies (i.e., policy trees), PI maintains a single controller for each agent and considers the value of beginning execution from any node of the controller. That is, starting at each (joint) node can be interpreted as an infinite-horizon (joint) policy and the set of policies can be considered to be the set of joint nodes. Pruning can then take place over nodes of these controllers to remove dominated policies.
2. Rather than initializing the iteration process with the (set of) one-stage-to-go policies as in DP, PI can start from any initial joint controller for each agent $m_0 = \langle m_{1,0}, \dots, m_{n,0} \rangle$. Exactly what this initial joint controller m_0 is does not matter: the controller is going to be subject to exhaustive backups, which will 'push' the initial controller to beyond the effective horizon (where the discount factor renders the value bounded by ε), as well as controller improvements.

© The Author(s) 2016
F.A. Oliehoek and C. Amato, *A Concise Introduction to Decentralized POMDPs*,
SpringerBriefs in Intelligent Systems, DOI 10.1007/978-3-319-28929-8_7

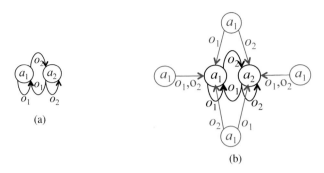

Fig. 7.1: A full backup for a single agent for action a_1 when starting with the controller in (a), resulting in the controller in (b).

The basic procedure of PI is that it continuously tries to improve the joint controller by improving the controller maintained for each agent. This per-agent improvement is done using an exhaustive backup operation (this is the same as in the finite-horizon DP case) by which nodes are added to the controller that intuitively correspond to all possible 'one-step longer' policies.[1] To counter the growth of the individual controllers, pruning can be conducted, which removes a node in an agent's controller if it is dominated (i.e., if it has equal or lower value than when beginning in a combination of nodes for all (s, I_{-i})-pairs). That is, the policy for agent i given by beginning in the node is dominated by some set of policies given by starting at other nodes in the controller. These exhaustive backups and pruning steps continue until the solution is provably within ε of an optimal solution. This algorithm can produce an ε-optimal policy in a finite number of steps [Bernstein et al., 2009]. The details of policy iteration follow.

The policy iteration algorithm is shown in Algorithm 7.1. The input is an initial joint controller, m_0, and a parameter ε. At each step, evaluation, backup and pruning occurs. The controller is evaluated using (6.3.1). Next, an exhaustive backup is performed to add nodes to each of the local controllers. Similarly to the finite-horizon case, for each agent i, $|\mathbb{A}_i||\mathbb{I}_i|^{|\mathbb{O}_i|}$ nodes are added to the local controller, where $|\mathbb{I}_i|$ is the number of nodes in the current controller. The exhaustive backup represents starting from each action and for each combination of observations transitioning to each of the nodes in the current controller. Repeated application of exhaustive backups amounts to a brute-force search in the space of deterministic policies, which will converge to ε-optimality, but is obviously quite inefficient.

To increase the efficiency of the algorithm, pruning takes place. Because planning takes place offline, the controllers for each agent are known at each step, but agents will not know which node of their controller any of the other agents will be in during execution. As a result, pruning must be completed over the multiagent belief

---

[1] Essentially, doing (infinitely many) exhaustive backups will generate all (infinitely many) policies.

Algorithm 7.1 Policy iteration for Dec-POMDPs.

Input: An initial joint controller $m_0 = \langle m_{1,0}, \ldots, m_{n,0} \rangle$.
Output: An ε-optimal joint controller m^*.
1: $\tau \leftarrow 0$
2: $m_\tau \leftarrow m_0$
3: **repeat**
4:    {Backup and evaluate:}
5:    **for** $i = 1$ to n **do**
6:       $m_{i,\tau} \leftarrow$ ExhaustiveBackup($m_{i,\tau-1}$)
7:    **end for**
8:    Compute V^{m_τ} {Evaluate the controllers}
9:    {Prune dominated nodes until none can be removed:}
10:   **while** some nodes have been pruned **do**
11:      **for** $i = 1$ to n **do**
12:         $m_{i,\tau} \leftarrow$ Prune($i, m_{-i,\tau}, m_{i,\tau}$)
13:         UpdateController($m_{i,\tau}$)   {Remove the pruned nodes and update links accordingly}
14:         Compute V^{m_τ} {Evaluate updated controllers}
15:      **end for**
16:   **end while**
17:   $\tau \leftarrow \tau + 1$
18: **until** $\frac{\gamma^{\tau+1}|R_{\max}|}{1-\gamma} \leq \varepsilon$
19: **return** $m^* \leftarrow m_\tau$

space (using a linear program that is very similar to that described for finite-horizon dynamic programming in Section 4.3). That is, a node for an agent's controller can only be pruned if there is some combination of nodes that has higher value for all states of the system and at all nodes of the other agents' controllers. Unlike in the finite-horizon case, edges to the removed node are then redirected to the dominating nodes. Because a node may be dominated by a distribution of other nodes, the resulting transitions may be stochastic rather than deterministic. The updated controller is evaluated, and pruning continues until no agent can remove any further nodes.

Convergence to ε-optimality can be calculated based on the discount rate and the number of iterations of backups that have been performed. Let $|R_{\max}|$ be the largest absolute value of any immediate reward in the Dec-POMDP. Then the algorithm terminates after iteration t if $\frac{\gamma^{t+1}|R_{\max}|}{1-\gamma} \leq \varepsilon$. At this point, due to discounting, the value of any policy after step t will be less than ε.

7.2 Optimizing Fixed-Size Controllers

Like optimal finite-horizon approaches, methods for producing ε-optimal infinite-horizon solutions are intractable for even moderately sized problems. This results in optimal algorithms not converging to any reasonable bound of the optimal solution in practice. To combat this intractability, approximate infinite-horizon algorithms

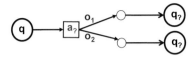

Fig. 7.2: Choosing actions and transitions in each node of a fixed-size controller.

have sought to produce a high-quality solution while keeping the controller sizes
for the agents fixed.

That is, the concept behind these approaches is to choose a controller size $|\mathbb{I}_i|$
for each agent and then determine action selection and node transitions parameters
that produce high values. However, because these methods fix the size of the agents'
controllers, the resulting size may be smaller than is needed for even an ε-optimal
solution. This means that, while some of the approaches in this section can produce a
controller that has the highest value *given* the fixed size, that value may be arbitrarily
far from the optimal solution.

7.2.1 Best-First Search

A way to compute the best *deterministic* joint controller of given size is via heuristic
search. One such method, referred to as *best-first search for Dec-POMDPs* [Szer and
Charpillet, 2005], generalizes the MAA* technique from Section 4.2.2 to the infinite
horizon. Rather than filling templates of trees (cf. Figure 5.1b) the method here fills
templates of controllers; it searches through the possible actions that can be taken
at each agent's controller nodes and the possible transitions that result from each
observation in each node.

In more detail, the algorithm searches through the possible deterministic transi-
tion mappings $\iota_i : \mathbb{I}_i \times \mathbb{O}_i \to \mathbb{I}_i$ and the deterministic action function, $\pi_i : \mathbb{I}_i \to \mathbb{A}_i$, for
all agents. The controller nodes are ordered and a forward search is conducted that
specifies the action selection and node transition parameters for all agents, one node
at a time. The heuristic value determines an upper bound value for these partially
specified controllers (again, in a way that is similar to MAA*) by assuming centralized control is used for unspecified nodes. In this way, an approximate value for the
controller is calculated given the currently specified deterministic parameters and
the algorithm fills in the remaining nodes one at a time in a best-first fashion.

This process continues until the value of a set of fully specified controllers is
greater than the heuristic value of any partially specified controllers. Since this is an
instance of heuristic search applied with an admissible heuristic, this technique is
guaranteed to find the best deterministic joint finite-state controller of a given size.

7.2.2 Bounded Policy Iteration

Because stochastic controllers with the same number of nodes are able to produce higher-valued policies than deterministic controllers, researchers have also explored optimizing stochastic controllers (a difficult problem by itself; see the overview and complexity results by Vlassis et al. 2012). These approaches seek to find probabilistic parameters for action selection and node transition. That is, for agent i, the algorithms find the probability of taking action a_i in node I_i, $\Pr(a_i|I_i)$, and the probability of transitioning to node I_i' in node I_i after action a_i is taken and observation o_i made, $\Pr(I_i'|I_i,a_i,o_i)$.

One method to produce stochastic controllers is by using a set of linear programs. In particular, *bounded policy iteration for decentralized POMDPs (Dec-BPI)* [Bernstein et al., 2005], is such a method that extends the BPI algorithm for POMDPs [Poupart and Boutilier, 2003] to the multiagent setting. This approach iterates through the nodes of each agent's controller, attempting to find an improvement for that node. That is, it tries to improve a single local controller, assuming that the controllers of the other agents are fixed, and thus is conceptually similar to JESP, described in Section 5.2.1. In contrast to JESP, however, this improvement for an agent i cannot be found using plain dynamic programming over a tree or directed acyclic graph (DAG) of reachable 'JESP beliefs' $b_i(\langle s, \bar{o}_{-i,t} \rangle)$ (even when replacing histories by internal states leading to a belief of the form $b_i(\langle s, I_{-i} \rangle)$ there would be infinitely many in general) or enumeration of all controllers m_i (there are uncountably many stochastic controllers). Instead, Dec-BPI uses linear programming to search for a new node to replace the old node.

Specifically, this approach iterates through agents i, along with nodes for agent I_i. Then, the method assumes that the current controller will be used from the second step on, and tries to replace the parameters for I_i with ones that will increase value for just the first step. That is, it attempts to find parameters satisfying the following inequality:

$$\forall s_t \in S, \forall I_{-i} \in \mathbb{I}_{-i}$$

$$V^m(s,I) \leq \sum_a \pi(a|I) \left[R(s,a) + \gamma \sum_{s,o',I'} \Pr(I'|I,a,o')\Pr(s',o'|s,a)V^m(s',I') \right].$$

Here, $\mathbb{I}_{-i}$ is the set of controller nodes for all agents besides i. The search for new parameters can be formulated as a linear program in Figure 7.3. Note that a 'combined' action selection and node transition probability,

$$\Pr(I_i',a_i|I_i,o_i) \triangleq \Pr(I_i'|I_i,a_i,o_i)\Pr(a_i|I_i),$$

is used to ensure the improvement constraints are linear; naive inclusion of the right-hand side product would lead to quadratic improvement constraints. Instead, we introduce more variables that lead to a linear formulation. The second probability constraint in Figure 7.3 ensures that the action selection probabilities can be recovered (i.e., that $y(a_i,o_i,I'_i)$ does not encode an invalid distribution). Note that the first and second probability constraints together guarantee that

$$\forall o_i \quad \sum_{I'_i,a_i} y(a_i,o_i,I'_i) = 1.$$

Given: I_i the node for which we want to test improvement.
variables:
ε, the value gap that we try to maximize ($\varepsilon > 0$ indicates an improvement),
$x(a_i) = \pi_i(a_i|I_i)$, action probability variables,
$y(a_i,o_i,I'_i) = \Pr(I'_i,a_i|I_i,o_i)$, combined action and next-stage node probabilities.
maximize: ε.
subject to:
Improvement constraints, $\forall s, I_{-i}$:

$$V^m(s,I) + \varepsilon \leq \sum_a \pi_{-i}(a_{-i}|I_{-i}) \left[x(a_i)R(s,a) + \gamma \sum_{s,o',I'} y(a_i,o_i,I'_i) \Pr(I'_{-i},s',o'|s,a)V^m(s',I') \right]$$

Probability constraints:

$$\sum_{a_i} x(a_i) = 1$$

$$\forall a_i,o_i \quad \sum_{I'_i} y(a_i,o_i,I'_i) = x(a_i)$$

$$\forall a_i \quad x(a_i) \geq 0$$

$$\forall a_i,o_i,I'_i \quad y(a_i,o_i,I'_i) \geq 0$$

Fig. 7.3: The linear program (LP) to test for improvement in Dec-BPI. The LP determines if there is a probability distribution over actions and transitions from node I_i that improves value when assuming the current controller will be used from the second step on. Note that $\Pr(I'_i,a_i|I_i,o_i)$ is the combined action and transition probability which is made consistent with the action selection probability $\pi_i(a_i|I_i)$ in the probability constraints. This form is needed to ensure the objective function is linear.

This linear program is polynomial in the sizes of the Dec-POMDP and the joint controller, but exponential in the number of agents. Bernstein et al. allow each agent's controller to be correlated by using shared information in a *correlation device* (as discussed in Section 6.2.4). This may improve solution quality while requiring only a limited increase in problem size [Bernstein et al., 2005, 2009].

7.2.3 Nonlinear Programming

Because Dec-BPI can often become stuck in local maxima, nonlinear programming (NLP) has also been used [Amato et al., 2007a, 2010]. The formulation seeks to optimize the value of a set of fixed-size controllers given an initial state distribution. The variables for this problem are the action selection and node transition probabilities for each node of each agent's controller as well as the value of the resulting policy starting from a set of controller nodes.

More formally, the NLP maximizes the expected value of the initial controller node for each agent at the initial belief subject to the Bellman constraints. To this end, let us translate the value of a joint controller (from Equation 6.3.1) in terms of variables that will be used for optimization:

$$z(I,s) = \sum_a \left[\prod_i x(I_i,a_i) \right]$$

$$\left(R(s,a) + \gamma \sum_{s',o} \Pr(s',o|s,a) \sum_{I'} \left[\prod_j y(I_j,a_i,o_i,I'_j) \right] z(I',s') \right). \quad (7.2.1)$$

As shown in Figure 7.4, $z(I,s)$ represents the value, $V(I,s_t)$, of executing the controller starting from nodes I and state s, while $x(I_i,a_i)$ is the action selection probability, $\Pr(a_i \mid I_i)$, and $y(I_i,a_i,o_i,I'_i)$ is the node transition probability, $\Pr(I'_i \mid I_i,a_i,o_i)$. Note that to ensure that the values are correct given the action and node transition probabilities, these nonlinear constraints must be added to the optimization which represent the Bellman equations given the policy determined by the action and transition probabilities. We must also ensure that the necessary variables are valid probabilities in a set of probability constraints. This approach differs from DEC-BPI in that it explicitly represents the node values as variables, thus allowing improvement and evaluation to take place simultaneously. An optimal solution of this nonlinear program represents an optimal fixed-size solution to the Dec-POMDP, but as this is often intractable, approximate solvers have been used in practice [Amato et al., 2010].

7.2.4 Expectation Maximization

As an alternative method for determining the parameters of stochastic controllers, expectation maximization (EM) has been used [Kumar and Zilberstein, 2010b, Kumar et al., 2011]. Again a fixed-size controller is assumed, but rather than determining the controller parameters using optimization, the problem is reformulated as a likelihood maximization problem and EM is used. This *planning as inference* technique is an extension of similar methods for POMDPs [Toussaint et al., 2006].

---

variables for each agent i :

$x(I_i,a_i) = \Pr(a_i \mid I_i)$, action probability variables,

$y(I_i,a_i,o_i,I_i') = \Pr(I_i' \mid I_i,a_i,o_i)$, next-stage node probabilities,

$z(I,s_t) = V(I,s_t)$, the values.

maximize:

$$\sum_{s_0} b_0(s_0)z(I_0,s_0).$$

subject to:

Bellman constraints:

$$\forall I,s \quad z(I,s) = \sum_a \left[\prod_i x(I_i,a_i) \right]$$

$$\left(R(s,a) + \gamma \sum_{s',o} \Pr(s',o|s,a) \sum_{I'} \left[\prod_j y(I_j,a_j,o_j,I_j') \right] z(I',s') \right). \quad (7.2.2)$$

Probability constraints for each agent i:

$$\forall I_i,a_i \quad \sum_{a_i} x(I_i,a_i) = 1$$

$$\forall I_i,a_i,o_i \quad \sum_{I_i'} y(I_i,a_i,o_i,I_i') = 1$$

$$\forall I_i,a_i \quad x(I_i,a_i) \geq 0$$

$$\forall I_i,a_i,o_i,I_i' \quad y(I_i,a_i,o_i,I_i') \geq 0$$

Fig. 7.4: The nonlinear program (NLP) representing the optimal fixed-size solution for the problem. The action selection, $\Pr(a_i|I_i)$, and node transition probabilities, $\Pr(I_i'|I_i,a_i,o_i)$, are optimized for each agent i to maximize the value of the controllers. This optimization is performed for the given initial belief b_0 and a given (arbitrarily selected) tuple of the initial nodes, $I_0 = \langle I_{1,0},\ldots,I_{n,0}\rangle$.

    The basic idea is that the problem can be represented as an infinite mixture of dynamic Bayesian networks (DBNs), which has one component for each time step t. The DBN responsible for a particular t covers stages $0,\ldots,t$ and represents the 'probability' that the 'maximum reward' is received at its last modeled stage (i.e., at t). The intuition is that the probability of achieving the maximum reward can be considered as a substitute for the value of the controller. We give a brief formalization of this intuition next; for details we refer the reader to the papers by Toussaint et al. [2006], Kumar and Zilberstein [2010b], and Kumar et al. [2011].

    First, the formalization is based on binary reward variables, r, for each stage t that provide probability via $\Pr(r = 1|s_t,a_t) \triangleq \frac{R(s_t,a_t)-R_{min}}{R_{max}-R_{min}}$, where R_{min} and R_{max} are the smallest and largest immediate rewards. This probability encodes the 'chance of getting the highest possible reward' at stage t. This can be used to define $\Pr(r,Z|t;\theta)$ with $Z = \langle s_0,a_0,s_1,o_1,I_1,a_1,\ldots,a_{t-1},s_t,o_t,I_t \rangle$ the entire histories of states, actions, observations and internal states, and with $\theta = \{\Pr(a|I),\Pr(I'|I,o),\Pr(I)\}$ the parameter vector that specifies the action selection, node transition and initial node probabilities. Now these probabilities constitute a mixture probability via:

$$\Pr(r,Z,t;\theta) = \Pr(r,Z|t;\theta)P(t),$$

with $P(t) = \gamma^t(1-\gamma)$ (used to discount the reward that is received for each time step t). This can be used to define an overall likelihood function $L^{\theta}(r=1;\theta)$, and it can be shown that maximizing this likelihood is equivalent to optimizing value for the Dec-POMDP (using a fixed-size controller). Specifically, the value function of controller θ can be recovered from the likelihood as $V(b_0) = \frac{(R_{max}-R_{min})L^{\theta}}{1-\gamma} + \sum_t \gamma^t R_{min}$ [Kumar and Zilberstein, 2010b].

As such, maximizing the value has been cast as the problem of maximizing likelihood in a DBN, and for this the EM algorithm can be used [Bishop, 2006]. It iterates by calculating occupancy probabilities—i.e., the probability of being in each controller node and problem state—and values given fixed controller parameters (in an E-step) and improving the controller parameters (in an M-step). The likelihood and associated value will increase at each iteration until the algorithm converges to a (possibly local) optima. The E-step calculates two quantities. The first is $P_t^{\theta}(I,s_t)$, the probability of being in state s_t and node I at each stage t. The second quantity, computed for each stage-to-go, is $P_{\tau}^{\theta}(r=1|I,s)$, which corresponds to the expected value of starting from I,s and continuing for τ steps. The M-step uses the probabilities calculated in the E-step and the previous controller parameters to update the action selection, node transition and initial node parameters.

After this EM approach was introduced, additional related methods were developed. These updated methods include EM for Dec-POMDPs with factored state spaces [Pajarinen and Peltonen, 2011a] and factored structured representations [Pajarinen and Peltonen, 2011b], and EM using simulation data rather than the full model [Wu et al., 2013].

7.2.5 Reduction to an NOMDP

Similarly to the transformation of finite-horizon Dec-POMDPs into NOMDPs described in Section 4.3, infinite-horizon Dec-POMDPs can also be transformed into (plan-time) NOMDPs. The basic idea here is to replace the observation histories by (a finite number of) information states such as nodes of an FSC. Let $I = \langle I_1,\ldots,I_n \rangle$ denote a joint information state. This allows us to redefine the plan-time sufficient statistic as follows:

$$\sigma_t(s,I) \triangleq \Pr(s,I|\delta_0,\ldots,\delta_{t-1}).$$

Again, this statistic can be updated using Bayes' rule. In particular $\sigma'(s',I')$ is given by

$$\forall_{(s',I')} \quad [U_{ss}(\sigma,\delta)](s',I') = \sum_{(s,I)} \Pr(s',I'|s,I,\delta(I))\sigma(s,I), \qquad (7.2.3)$$

where—using $\iota(I'|I,a,o) = \prod_{i\in\mathbb{D}} \iota_i(I'_i|I_i,a_i,o_i)$ for the joint information state update probability—the probability of transitioning to (s',I') is given by

$$\Pr(s',I'|s,I,a) = \Pr(s'|s,a) \sum_o \iota(I'|I,a,o) \Pr(o|a,s'). \qquad (7.2.4)$$

It is easy to show that, *for a given set $\{\iota_i\}$ of information state functions*, one can construct a plan-time NOMDP analogous to Definition 23 in Section 4.3.3, where augmented states are tuples $\bar{s} = \langle s,I\rangle$. However, as discussed before, in the infinite-horizon setting, the selection of those information state functions becomes part of the problem.

One idea to address this, dating back to Meuleau et al. [1999a], is to make searching the space of deterministic information state functions part of the problem by defining a cross-product MDP in which "a decision is the choice of an action and of a next node". That is, selection of the ι_i function (in a POMDP with protagonist agent i) can be done by introducing $|\mathbb{O}_i|$ new action variables (say, $a_i^\iota = \{a_i^{\iota,1},\ldots,a_i^{\iota,|\mathbb{O}_i|}\}$) that specify, for each observation $o_i \in \mathbb{O}_i$, to what next internal state to transition. This approach is extended to Dec-POMDPs by Mac-Dermed and Isbell [2013] who introduce the *bounded belief Dec-POMDP*[2] *(BB-Dec-POMDP)*, which is a Dec-POMDP that encodes the selection of optimal $\{\iota_i\}$ by splitting each stage into two stages: one for selection of the domain-level actions and one for selection of the a_i^ι. We omit the details of this formulation and refer the reader to the original paper. The main point that the reader should note is that by making ι part of the (augmented) joint action, the probability $\Pr(s',I'|s,I,a)$ from (7.2.4) no longer depends on external quantities, which means that *it is possible to construct an NOMDP formulation analogous to Definition 23 that in fact does optimize over (deterministic) information state functions.*

This is in fact what MacDermed and Isbell [2013] propose; they construct the NOMDP (to which they refer as a 'belief-POMDP') for a BB-Dec-POMDP and solve it with a POMDP solution method. Specifically, they use a modification of the point-based method Perseus [Spaan and Vlassis, 2005] to solve the NOMDP. The modification employed is aimed at mitigating the bottleneck of maximizing over (exponentially many) decision rules in $V^*(\sigma) = \max_\delta Q^*(\sigma,\delta)$. Since the value function is PWLC, the next-stage value function can be represented using a set of vectors $v \in \mathscr{V}$, and we can write

$$V^*(\sigma) = \max_\delta \sum_{(s,I)} \sigma(s,I) \left(R(s,\delta(I)) + \max_{v\in\mathscr{V}} \sum_{(s',I')} \Pr(s',I'|s,I,\delta) v(s',I') \right)$$

$$= \max_{v\in\mathscr{V}} \max_\delta \sum_{(s,I)} \sigma(s,I) \underbrace{\left(R(s,\delta(I)) + \sum_{(s',I')} \Pr(s',I'|s,\delta(I)) v(s',I') \right)}_{v_\delta(s,I)}.$$

---

[2] The term 'bounded belief' refers to the finite number of internal states (or 'beliefs') considered.

The key insight is the in the last expression, the bracketed part only depends on $\delta(I) = \langle \delta_1(I_1), \ldots, \delta_1(I_1) \rangle$, i.e., on that part of δ that specifies the actions for I only. As such it is possible to rewrite this value as the maximum of solutions of a collection of collaborative Bayesian games (cf. Section 5.2.3), one for each $v \in \mathcal{V}$:

$$V^*(\sigma) = \max_{v \in \mathcal{V}} \max_{\delta} \sum_I \sigma(I) \sum_s \sigma(s|I) v_\delta(s,I)$$

$$= \max_{v \in \mathcal{V}} \left[\max_{\delta} \sum_I \sigma(I) Q^v(I, \delta(I)) \right].$$

For each $v \in \mathcal{V}$, the maximization over δ can be interpreted as the solution of a CBG (5.2.2), and therefore can be performed more effectively using a variety of methods [Oliehoek et al., 2010, Kumar and Zilberstein, 2010a, Oliehoek et al., 2012a]. MacDermed and Isbell [2013] propose a method based on the relaxation of an integer program. We note that the maximizing δ directly induces a vector v_δ, which is the result of the point-based backup. As such, this modification can also be used by other point-based POMDP methods.

It is good to note that a BB-Dec-POMDP is just a special case of an (infinite-horizon) Dec-POMDP. The fact that it happens to have a bounded number of information states is nothing new compared to previous approaches: those also limited the number of information states (controller nodes) to a finite number. The conceptual difference, however, is that MacDermed and Isbell [2013] pose this restriction as part of the *model definition*, rather the solution method. This is very much in line with, and a source of inspiration for, the more general definition of multiagent decision problems that we introduced in Section 2.4.4.

Chapter 8
Further Topics

Decentralized decision making is an active field of research, and many different directions and aspects are actively explored. Here we present an overview of some of the more active directions that people have been pursuing in recent years.

8.1 Exploiting Structure in Factored Models

One of the major directions of research in the last decade has been the identification and exploitation of structure in Dec-POMDPs. In particular, much research has considered the special cases mentioned in Section 2.4.2, and models that generalize these.

8.1.1 Exploiting Constraint Optimization Methods

One of the main directions of work that people have pursued is the exploitation of structure between variables by making use of methods from constraint optimization or inference in graphical models. These methods construct what is called a coordination graph, which indicates the agents that need to directly coordinate. Methods to exploit the sparsity of this graph can be employed for more efficient computation. After introducing coordination graphs, this section will give an impression of how they are employed within ND-POMDPs and factored Dec-POMDPs.

8.1.1.1 Coordination (Hyper-)Graphs

The main idea behind *coordination graphs* [Guestrin et al., 2002a, Kok and Vlassis, 2006], also called *interaction graphs* [Nair et al., 2005] or *collaborative graphical games* [Oliehoek et al., 2008c, 2012a], is that even though all agents are relevant

© The Author(s) 2016
F.A. Oliehoek and C. Amato, *A Concise Introduction to Decentralized POMDPs*,
SpringerBriefs in Intelligent Systems, DOI 10.1007/978-3-319-28929-8_8

for the total payoff, not all agents directly interact with each other. For instance, Figure 8.1a shows a coordination graph involving four agents. All agents try to optimize the sum of the payoff functions, but agent 1 only *directly* interacts with agent 2. This leads to a form of *conditional* independence that can be exploited using a variety of techniques.

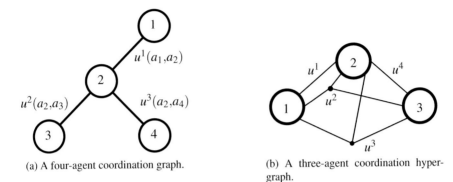

(a) A four-agent coordination graph.   (b) A three-agent coordination hyper-graph.

Fig. 8.1: Example Coordination Graphs.

Definition 28 (Coordination (Hyper-)Graph). A *coordination (hyper-)graph (CG)* is a tuple $CG = \langle \mathbb{D}, \mathbb{A}, \{u^1, \ldots, u^\rho\} \rangle$, where

- $\mathbb{D} = \{1, \ldots, n\}$ is the set of n agents.
- $\mathbb{A}$ is the set of joint actions.
- $\{u^1, \ldots, u^\rho\}$ is a set of *local* payoff functions.

Each local payoff function u^e is a mapping of the actions of a subset of agents $e \subseteq \mathbb{D}$ to payoffs, called the *scope* of the function. Note that we use e to denote both the index and the scope of the payoff function, such that we write $u^e(a_e)$.[1] As illustrated in Figure 8.1, the scopes of the local payoff function induce a (hyper-)graph: there is a set of (hyper-)edges $\mathscr{E}$ such that there is one edge $e \in \mathscr{E}$ for every local payoff function. Each edge connects one, two or more agents (nodes) $e = \{i, j, \ldots\} \subseteq \mathbb{D}$. So the graph is a traditional graph when exactly two agents participate in each local payoff function and a hyper-graph otherwise.

The goal in a CG is to find the joint action with the highest total payoff:

$$u(a) = \sum_{e \in \mathscr{E}} u^e(a_e).$$

---

[1] Even though an abuse of notation, the meaning should be clear from context. Moreover, it allows us to avoid the notational burden of correct alternatives such as defining the subset as $\mathscr{A}(e)$ and writing $u^e(a_{\mathscr{A}(e)})$.

Note that the local reward functions should not be confused with *individual* reward functions in a system with self-interested agents, such as partially observable stochastic games [Hansen et al., 2004] and (graphical) BGs [Kearns et al., 2001, Kearns, 2007]. In such models agents compete to maximize their individual reward functions, while we consider agents that collaborate to maximize the sum of local payoff functions.

A CG is a specific instantiation of a *constraint optimization problem (COP)* [Marinescu and Dechter, 2009] or, more generally, a *graphical model* [Dechter, 2013] where the nodes are agents (rather than variables) and they can take actions (rather than values). This means that there are a multitude of methods that can be brought to bear on these problems. The most popular exact method is *nonserial dynamic programming (NDP)* [Bertele and Brioschi, 1972], also called *variable elimination* [Guestrin et al., 2002a, Kok and Vlassis, 2006] and *bucket elimination* [Dechter, 1997]. Since NDP can be slow for certain network topologies (those with high 'induced width'; see Kok and Vlassis 2006 for details and empirical investigation), a popular (but approximate) alternative is the max-sum algorithm [Pearl, 1988, Kok and Vlassis, 2005, 2006, Farinelli et al., 2008, Rogers et al., 2011]. It is also possible to use distributed methods as investigated in the field of *distributed constraint optimization problems (DCOPs)* [Liu and Sycara, 1995, Yokoo, 2001, Modi et al., 2005].

The framework of CGs can be applicable to Dec-POMDPs when replacing CG-actions by Dec-POMDP policies. That is, we can state a requirement that the value function of a Dec-POMDP can be additively decomposed into local values:

$$V(\pi) = \sum_{e \in \mathcal{E}} V^e(\pi_e), \tag{8.1.1}$$

where $\pi_e = \langle \pi_{e1}, \ldots, \pi_{e|e|} \rangle$ is the profile of individual policies of agents that participate in edge e. The class of Dec-POMDPs that satisfy this requirement, also called *value-factored Dec-POMDPs* [Kumar et al., 2011], can be trivially transformed to CGs.

8.1.1.2 ND-POMDPs

Nair et al. [2005] introduced the *networked distributed POMDP (ND-POMDP)*, which is a subclass of Dec-POMDPs that imposes constraints such that 8.1.1 holds. An ND-POMDP can be described as a transition- and observation-independent Dec-POMDP with additively separated rewards.

Definition 29. An *ND-POMDP* is a Dec-POMDP, where

- The states are agent-wise factored: $\mathbb{S} = \mathbb{S}_0 \times \mathbb{S}_1 \times \ldots \times \mathbb{S}_n$ (as discussed in Section 2.4.2).
- The reward function exhibits *additive separability,* which means that it is possible to decompose the reward function into the sum of smaller *local* reward functions:

$$R(s,a) = \sum_{e=1}^{\rho} R^e(s_0, s_e, a_e). \tag{8.1.2}$$

Again, e is overloaded to also denote a subset of agents.
- The transition and observation model satisfy the transition and observation independence properties from Section 2.4.2.

Note that the ND-POMDP allows for an additional state component that takes values $s_0 \in \mathbb{S}_0$, and that can be used to model aspects of the world that cannot be influenced by the agents, such as the weather. This means that its transitions only depend on itself, $T_i(s_0'|s_0)$. However, the non-influenceable state factor is allowed to influence the observation of the agents: $O_i(o_i|a_i, s_i', s_0')$, as well as the reward components (8.1.2).

Nair et al. [2005] show that in this setting the value can be decomposed into local value functions: i.e., (8.1.1) holds. This means that the factorization of the value function implies a coordination (hyper-)graph, and that (distributed) COP methods can be employed: in each of the nodes (representing agents) a policy is chosen, and the goal is to select a joint policy such that the sum of local value functions is maximized. The sparsity in the graph can be exploited for more efficient computation.

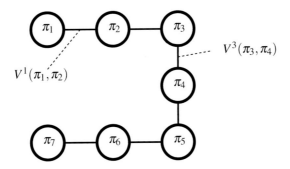

Fig. 8.2: An ND-POMDP sensor network example.

Typical motivating domains for ND-POMDPs include sensor network and target tracking problems. Figure 8.2 shows an example coordination graph corresponding to an ND-POMDP for the sensor network problem from Section 2.3 (see Figure 2.7). There are six edges, each connecting two agents, which means that the reward function decomposes as follows:

$$R(s,a) = R^1(s_0, s_1, s_2, a_1, a_2) + R^2(s_0, s_2, s_3, a_2, a_3) + \cdots + R^6(s_0, s_6, s_7, a_6, a_7).$$

Let $\mathcal{N}_i$ denote the neighbors of agent i, including i itself, in the interaction graph. In an ND-POMDP, the *local neighborhood utility* of an agent i is the expected return for all the edges that contain agent i:

$$V(\pi_{\mathcal{N}_i}) = \sum_{e \in \mathcal{E} \text{ s.t. } i \in e} V^e(\pi_e). \tag{8.1.3}$$

It can be shown that when an agent $j \notin \mathcal{N}_i$ changes its policy, $V(\pi_{\mathcal{N}_i})$ is not affected, a property referred to as *locality of interaction* [Nair et al., 2005].

This locality of interaction is the crucial property that allows COP methods to optimize more effectively.[2] In particular, Nair et al. [2005] propose a *globally optimal algorithm (GOA)*, which essentially performs nonserial dynamic programming on the interaction graph, as well as *locally interacting distributed joint equilibrium search for policies (LID-JESP)*, which combines the *distributed breakout algorithm (DBA)* [Yokoo and Hirayama, 1996] with JESP (see Section 5.2.1). Kim et al. [2006] extend this method to make use of a stochastic version of DBA, allowing for a speedup in convergence. Both Varakantham et al. [2007b] and Marecki et al. [2008] address the COP using heuristic search (for the finite and the infinite-horizon case respectively). Kumar and Zilberstein [2009] extend MBDP (see Section 5.2.2) to exploit COP methods in both the heuristic used to sample belief points and in computing the best subtree for a sampled belief. Kumar et al. [2011] cast the planning problem as an inference problem [Toussaint, 2009] and employ the *expectation maximization (EM)* algorithm (e.g., see Koller and Friedman, 2009) to solve the problem. Effectively, this approach decomposes the full inference problem into smaller problems by using message passing to compute the local values (E-step) of each factor and then combining the resulting solutions in the M-step. Dibangoye et al. [2014] use the reduction to an NOMDP (discussed in Section 4.3) and extend the solution method of Dibangoye et al. [2013] by exploiting the factored structure in the resulting NOMDP value function.

We note that even with factored transitions and observations, a policy in an ND-POMDP is a mapping from observation histories to actions (unlike in the transition- and observation-independent Dec-MDP case, where policies are mappings from local states to actions) and the worst-case complexity remains the same as in a regular Dec-POMDP (NEXP-complete), thus implying doubly exponential complexity in the horizon of the problem. While the worst-case complexity remains the same as in the Dec-POMDP case, algorithms for ND-POMDPs are typically much more scalable *in the number of agents* in practice. Scalability can increase as the hyper-graph becomes less connected.

8.1.1.3 Factored Dec-POMDPs

The ND-POMDP framework enables the exploitation of locality by imposing restrictions that guarantee that the value function factorizes in local components. Unfortunately, these restrictions narrow down the class of problems that can be represented. Alternatively, we can investigate a broader class and see to what extent locality of interaction might still hold.

---

[2] The formula (8.1.3), that explains the term *locality of interaction*, and (8.1.1), that states the requirement that the *value function can be decomposed into local components*, each emphasize different aspects. Note, however that one cannot exist without the other and that, therefore, these terms can be regarded as synonymous.

Definition 30 (Factored Dec-POMDP). A *factored Dec-POMDP (fDec-POMDP)* [Oliehoek et al., 2008c] is identical to a Dec-POMDP with:

- an arbitrary (non-agent-wise) factored state space $\mathbb{S} = \mathbb{X}^1 \times \ldots \times \mathbb{X}^{|\mathcal{X}|}$, which is spanned by a set of factors $\mathbb{X} = \{\mathbb{X}^1, \ldots, \mathbb{X}^{|\mathcal{X}|}\}$;
- a reward function that exhibits *additive separability*: $R(s,a) = \sum_{e=1}^{\rho} R^e(x_e, a_e)$, which is similar to the above definition in (8.1.2), but now x_e expresses the value of an arbitrary subset of state factors associated with reward component e.

In a factored Dec-POMDP, the transition and observation model can be compactly represented by exploiting conditional independence between variables. In particular, the transition and observation model can be represented by a dynamic Bayesian network (DBN) [Boutilier et al., 1999]. In such a DBN, arrows represent causal influence and each node with incoming edges has a conditional probability table (CPT) associated with it. Although the size of these CPTs is exponential in the number of parents, the parents are typically a subset of all state factors and actions, leading to a model exponentially smaller than a flat model. Rewards can be included in the DBN.[3] The reward nodes have conditional reward tables (CRTs) associated with them that represent the local reward functions. Decision trees [Boutilier et al., 1999, 2000] or algebraic decision diagrams [Bahar et al., 1993, St-Aubin et al., 2001, Poupart, 2005] can be used to further reduce the size of representation of both CPTs and CRTs by exploiting *context-specific independence*. For example, the value of some factor x^i may be of no influence when some other factor x^j has a particular value.

As an example, we consider the FIREFIGHTINGGRAPH (FFG) problem, illustrated in Figure 8.3a. It models a team of n firefighters that have to extinguish fires in a row of $n_H = n + 1$ houses. In the illustration, we assume $n = 3, n_H = 4$. At every time step, each agent i can choose to fight fires at house i or $i + 1$. For instance, agent 2's possible actions are H_2 and H_3. Each agent can observe whether there are flames, $o_i = F$, or not, $o_i = N$, at its location. Each house H is characterized by a fire level x^H, an integer parameter in $[0, N_f)$, where a level of 0 indicates the house is not burning. A state in FFG is an assignment of fire levels $s = \langle x^1, x^2, x^3, x^4 \rangle$. Initially, the fire level x^H of each house is drawn from a uniform distribution.

Figure 8.3b shows the DBN for the problem. The transition probabilities are such that a burning house where no firefighter is present has a reasonable chance to increase its fire level. When a neighboring house is also burning, this probability is even higher. In addition a burning neighboring house can also ignite a non-burning house. When two agents are in the same house, they will extinguish any present fire completely, setting the house's fire level to 0. A single agent present at a house will lower the fire level by one point with probability 1 if no neighbors are burning, and with probability 0.6 otherwise. Flames are observed with probability 0.2 if the visited house is not burning, with probability 0.5 if its fire level is 1, and with probability 0.8 otherwise.

---

[3] Such a DBN that includes reward nodes is also called an *influence diagram (ID)* [Howard and Matheson, 1984].

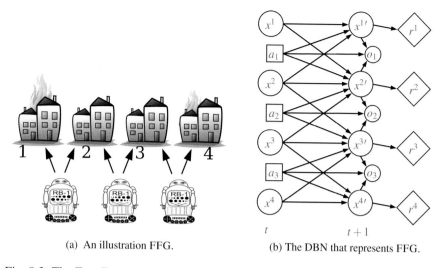

(a) An illustration FFG.   (b) The DBN that represents FFG.

Fig. 8.3: The FIREFIGHTINGGRAPH problem. A DBN can be used to represent the transition, observation and reward function.

Figure 8.3b also includes the local reward functions, one per house. In particular, for each house $1 \leq H \leq 4$ the rewards are specified by the fire levels at the next time step $r^H(x^{H\prime}) = -x^{H\prime}$. We can reduce these rewards to ones of the form $R^H(s,a)$ as in Definition 2 by taking the expectation over $x^{H\prime}$. For instance, for house 1

$$R^1(x_{\{1,2\}},a_1) = \sum_{x^{1\prime}} \Pr(x^{1\prime}|x_{\{1,2\}},a_1)r^1(x^{1\prime}), \qquad (8.1.4)$$

where $x_{\{1,2\}}$ denotes $\langle x^1, x^2 \rangle$. While FFG is a stylized example, such locally connected systems can be found in applications as traffic control [Wu et al., 2013] or communication networks [Ooi and Wornell, 1996, Hansen et al., 2004, Mahajan and Mannan, 2014].

Value Functions for Factored Dec-POMDPs   Factored Dec-POMDPs can exhibit very local behavior. For instance, in FFG, the rewards for each house only depend on a few local factors (both 'nearby' agents and state factors), so it seems reasonable to expect that also in these settings we can exploit locality of interaction. It turns out, however, that due to the absence of transition and observation independence, the story for fDec-POMDPs is more subtle than for ND-POMDPs. In particular, in this section we will illustrate this by examining how the Q-function for a particular joint policy Q^π decomposes.

Similarly to typical use in MDPs and RL, the Q-function is defined as the value of taking a joint action and following π subsequently (cf. the definition of V^π by (3.4.2)):

$$Q^{\pi}(s_t,\bar{o}_t,a_t) \triangleq \begin{cases} R(s_t,a_t), \text{ for the last stage } t = h-1, \text{ and otherwise:} \\ R(s_t,a_t) + \sum_{s_{t+1}} \sum_{o_{t+1}} \Pr(s_{t+1},o_{t+1}|s_t,a_t) Q^{\pi}(s_{t+1},\bar{o}_{t+1},\pi(\bar{o}_{t+1})). \end{cases}$$

Since the immediate reward function is additively factored in case of an fDec-POMDP, it is possible to decompose this Q-function into ρ components Q^e that each predict the value for a reward component R^e. This decomposition is not as straightforward as the decomposition of values in ND-POMDPs (8.1.1), but needs to deal with differences in structure between stages, as we explain next.

For the last stage of an fDec-POMDP, the Q-function is trivially additively factored: Q_{h-1}^{π} is simply equal to the immediate reward function. This is illustrated for FFG in Figure 8.4. The figure shows that the Q-function can be decomposed in four local Q-value functions $Q = Q^1 + \cdots + Q^4$, where each Q^e is defined over the same subset of variables as R^e (the *scope* of Q^e). In order to exploit independence between agents, we discriminate between the variables that pertain to state factors and those that pertain to agents (i.e., actions and observations).

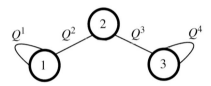

Fig. 8.4: The interaction hyper-graph illustrating the factorization of Q^{π} for FIRE-FIGHTINGGRAPH $t = h-1$.

For other stages, $t < h-1$, it turns out that—even without assuming transition and observation independence—it is still possible to decompose the value function as the sum of functions, one for each payoff component e. However, because actions, factors and observations influence each other, the scopes of the components Q^e grow over time. This is illustrated in Figure 8.5, which shows the scope of Q^1, the value component that represents the expected (R^1) reward for house 1 at different stages of a horizon $h = 3$ FFG problem. Even though the scope of the function R^1 only contains $\{x^1,x^2,a_1\}$, at earlier stages we need to include more variables since they can affect R_{h-1}^1, the reward for house 1 at the last stage $t = h-1$.

We can formalize these growing scopes using 'scope backup operators' for state factors scopes ($\Gamma^{\mathscr{X}}$) and agent scopes ($\Gamma^{\mathscr{A}}$). Given a set of variables $\mathbb{V}$ (which can be either state factors or observations) from the right-hand side of the 2DBN, they return the variables (respectively the state factors scope and agent scope) from the left-hand side of the 2DBN are ancestors of $\mathbb{V}$. This way, it is possible to specify a local value function component:

$$Q_{\pi}^e(x_{e,t},\bar{o}_{e,t},a_e) = R^e(x_{e,t},a_e) + \sum_{x_{e',t+1}} \sum_{o_{e',t+1}}$$

$$\Pr(x_{e',t+1},o_{e',t+1}|x_{\Gamma^{\mathscr{X}},t},a_{\Gamma^{\mathscr{A}}}) Q_{\pi}^e(x_{e',t+1},\bar{o}_{e',t+1},\pi_{e'}(\bar{o}_{e',t+1})). \quad (8.1.5)$$

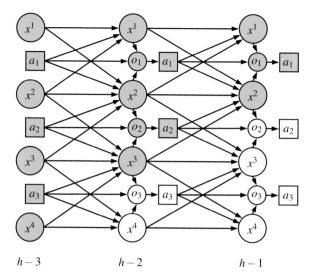

Fig. 8.5: Illustration of the interaction between agents and environment over time in FFG. In contrast to Figure 8.3b, which represents the transition and observation model using abstract time steps t and $t+1$, this figure represents the last 3 stages of a decision problem. Also the rewards are omitted in this figure. The scope of Q^1, given by (8.1.4), is illustrated by shading and increases when going back in time.

where $x_{\Gamma\mathscr{X},t}, a_{\Gamma\mathscr{A}}$ denote the state factors and actions of agents that are needed to predict the state factors ($x_{e',t+1}$) as well as the observations of the agents ($o_{e',t+1}$) in the next stage scope.

Theorem 3 (Decomposition of $V^t(\pi)$). *Given an additively factored immediate reward function, for any t there is a specification of scopes $(x_{e,t}, \bar{o}_{e,t})$ such that the value $V_t(\pi)$ of a finite-horizon factored Dec-POMDP is decomposable:*

$$V_t(\pi) = \sum_{e\in\mathscr{E}} V_t^e(\pi) = \sum_{e\in\mathscr{E}}\sum_{x_{e,t}}\sum_{\bar{o}_{e,t}}\Pr(x_{e,t},\bar{o}_{e,t}|b_0,\pi)Q_\pi^e(x_{e,t},\bar{o}_{e,t},\pi_e(\bar{o}_{e,t})). \quad (8.1.6)$$

Proof. See Oliehoek [2010, Chapter 5].

When one or more components becomes 'fully coupled' (i.e., contains all agents and state factors), technically, the value function still is additively factored. However, at this point the components are no longer local, which means that factorization will no longer provide any benefits. Therefore, at this point the components can be collapsed to form a non-factored value function.

When assuming transition and observation independence, as in ND-POMDPs, the scopes do not grow: each variable $s_{i,t+1}$ (representing a local state for agent i) is dependent only on $s_{i,t}$ (its own value at the previous state), and each o_i is dependent only on $s_{i,t+1}$. As a result, the interaction graph for such settings is stationary.

In the more general case, such a notion of locality of interaction over full-length policies is not properly defined, because the interaction graph and hence an agent's neighborhood can be different at every stage. It is possible to define such a notion for each particular stage t [Oliehoek et al., 2008c]. As a result, it is in theory possible to exploit factorization in an exact manner. However, in practice it is likely that the gains of such a method will be limited by the dense coupling for earlier stages.

A perhaps more promising idea is to exploit factored structure in approximate algorithms. For instance, even though Figure 8.5 clearly shows there is a path from x^4 to the immediate reward associated with house 1, it might be the case that this influence is only minor, in which case it might be possible to approximate Q^1 with a function with a smaller scope. This is the idea behind *transfer planning* [Oliehoek et al., 2013b]: the value functions of abstracted source problems involving few state factors and agents are used as the component value functions for a larger task. In order to solve the constraint optimization problem more effectively, the approach makes use of specific constraint formulation for settings with imperfect information [Oliehoek et al., 2012a]. Other examples are the methods by Pajarinen and Peltonen [2011a] and Wu et al. [2013] that extend the EM method for Dec-POMDPs by Kumar and Zilberstein [2010b] and Kumar et al. [2011] such that they can be applied to fDec-POMDPs. While these methods do not have guarantees, they can be accompanied by methods to compute upper bounds for fDec-POMDPs, such that it is still possible to get some information about their quality [Oliehoek et al., 2015a].

8.1.2 Exploiting Influence-Based Policy Abstraction

The reduction of (special cases of) Dec-POMDP to constraint optimization problems, as described above, exploits *conditional independence of the best responses of agents*. That is, the best response of one agent might depend only on a subset of other agents. However, in certain (special cases of) factored Dec-POMDPs there is also another type of structure that has been exploited [Becker et al., 2003, 2004a, Varakantham et al., 2009, Petrik and Zilberstein, 2009, Witwicki and Durfee, 2010a,b, 2011, Velagapudi et al., 2011, Witwicki, 2011, Witwicki et al., 2012, Oliehoek et al., 2012b]. This type of structure, which we refer to as *compact influence space*, operates at a finer level and may even be exploited in settings with just two agents.

The core idea is that in an interaction, say between a cook i and a waiter j, not all details of the agents' policies are relevant for their interaction. There might be many different ways for the cook to prepare a steak, but the only thing that is relevant for the policy of the waiter is *when* the steak is ready to be served. As such, for the waiter to compute a best response to the cook's policy π_i, he does not need to know every detail about that policy, but instead can compute a more abstract representation of that policy, to which we will refer as the *(experienced) influence* $I_{i \rightarrow j}$ of policy π_i on agent j.

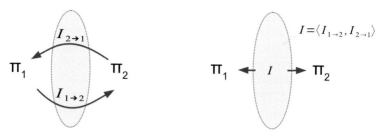

(a) IBA: for the purpose of computing best responses, the policies of each agent can be represented in a more abstract way as influences.

(b) IS: given a joint influence point, each agent i can independently compute its best-response (subject to the constraint of satisfying $I_{i \to j}$).

Fig. 8.6: Influence-based abstraction (IBA) and influence space search (IS).

The process of creating an abstract representation of π_i is called *influence-based (policy) abstraction (IBA)*, and is illustrated in Figure 8.6a. A number of different approaches have been proposed for performing this operation [e.g., Becker et al., 2003, Varakantham et al., 2009, Witwicki and Durfee, 2010a, Oliehoek et al., 2012b], exploiting different properties of the specific model for which they were proposed. In the case that IBA can be performed efficiently, it can be useful for best-response computations, with many potential applications. From the perspective of objective planning for Dec-POMDPs, however, there is another very important application: since the IBA operator potentially maps many policies π_i to the same influence $I_{i \to j}$, the so-called *(joint) influence space* can be much smaller than the space of joint policies. This means that searching the space of joint influences may lead to substantial improvements in efficiency. This idea of joint influence space search is illustrated in Figure 8.6b. As the figure emphasizes, influence search is not only a promising technique because the the space of joint influences may be smaller, but also because it allows each agent to compute its (constrained) best response locally, thus opening the door to distributed *planning* approaches.

As a concrete example Becker et al. [2004b] consider a domain in which Mars rovers need to perform exploration missions. Each rover has its own mission where it visits a number of sites to take pictures, conduct experiments, collect soil samples, etc. During the execution of the mission of a rover, unexpected events (delays) can occur, which means that at every site the rover needs to decide whether to perform the planned action or to continue to the next site. While the execution of the missions do not interfere with each other (i.e., there is transition and observation independence; cf. Section 2.4.2), the rewards do: some of the rovers may visit the same site and the joint reward can be sub- or superadditive depending on the task. For instance, consider a scenario where we want to find a joint policy for two rovers i and j that both visit different sides of a canyon. Since two pictures from the different sides of the canyon will give an impression of the whole canyon, this might be worth a lot more than twice the utility of a single picture. This means that the best response for agent i depends on the probability that π_j will lead to agent j successfully taking a picture from his side of the canyon. That is, the probability $p_j(foto|\pi_j)$ that j successfully takes the picture is $I_{j \to i}$, the *influence* that π_j exerts on agent i, and similarly $p_i(foto|\pi_i)$ corresponds to $I_{i \to j}$. To find optimal solutions for such problems, Becker et al. [2003] introduce the *coverage-set algorithm*, which searches over the space of probabilities p_j (not unlike the linear support algorithm for POMDPs by Cheng [1988]) and computes all possible best response policies for agent i. The resulting set, dubbed the *coverage set*, is subsequently used to compute all the candidate joint policies (by computing the best responses of agent j) from which the best joint policy is selected.

The above example illustrates the core idea of influence search. Since then, research has considered widening the class of problems to which this approach can be applied [Becker et al., 2004a, Varakantham et al., 2009, Witwicki and Durfee, 2010b, Velagapudi et al., 2011, Oliehoek et al., 2012b], leading to different definitions of 'influence' as well as different ways of performing the influence search [Varakantham et al., 2009, Petrik and Zilberstein, 2009, Witwicki et al., 2012, Oliehoek et al., 2012b]. Currently, the broadest class of problems for which influence search has been defined is the so-called *transition-decoupled POMDP (TD-POMDP)* [Witwicki and Durfee, 2010b], while IBA has also been defined for factored POSGs [Oliehoek et al., 2012b]. Finally, the concept of influence-based abstraction and influence search is conceptually similar to techniques that exploit *behavioral equivalence* in subjective (e.g., I-POMDP, cf. Section 2.4.6) planning approaches [Pynadath and Marsella, 2007, Rathnasabapathy et al., 2006, Zeng et al., 2011]; the difference is that these approaches abstract classes of behavior down to policies, whereas IBA abstracts policies down to even more abstract influences.

8.2 Hierarchical Approaches and Macro-Actions

The Dec-POMDP framework requires synchronous decision making: every agent determines an action to execute, and then executes it in a single time step. While this does not rule out things like turn-taking—it would be easy to force an agent to select a 'noop' action based on an observable bit, for instance—this restriction is problematic in settings with variable-length action duration and (thus) in hierarchical approaches [Oliehoek and Visser, 2006, Messias, 2014, Amato et al., 2014].

Many real-world systems have a set of controllers (e.g, way point navigation, grasping an object, waiting for a signal), and planning consists of sequencing the execution of those controllers. These controllers are likely to require different amounts of time, so synchronous decision making would require waiting until all agents have completed their controller execution (and achieved common knowledge of this fact).

Similarly, such delays could, and typically would, arise in hierarchical solutions to Dec-POMDPs. For instance, Oliehoek and Visser [2006] describe an idea for a hierarchical approach to addressing RoboCup Rescue: the lowest level corresponds to a regular Dec-POMDP, while higher levels correspond to assignments of agents to particular fires or regions. Again, the problem here is that it is not possible guarantee that the tasks 'fight fire A' and 'fight fire B' taken by two subsets of agents will end simultaneously, and a synchronized higher level will therefore induce delays: the amount of time Δt needs to be long enough to guarantee that all tasks are completed, and if a task finishes sooner, agents will have to wait. In some cases, it may be possible to overcome the worst effects of such delays by expanding the state space (e.g., by representing how far the current task has progressed, in combination with time step lengths Δt that are not too long). In many cases, however, more principled methods are needed: the differences in task length may make the above approach cumbersome, and there are no guarantees for the loss in quality it may lead to. Moreover, in some domains (e.g., when controlling airplanes or underwater vehicles that cannot stay in place) it is not possible to tolerate *any* delays.

One potential solution for MPOMDPs, i.e., settings where free communication is available, is provided by Messias [2014], who describes multirobot decision making as an event-driven process: in the Event-Driven MPOMDP framework, every time that an agent finishes its task, an *event* occurs. This event is immediately broadcast to other agents, who become aware of the new state of the environment and thus can immediately react. Another advantage of this approach is that, since events are never simultaneous (they occur in continuous time), this means that the model does not suffer from exponentially many joint observations. A practical difficulty for many settings, however, is that the approach crucially depends on free and instantaneous broadcast communication.

A recent approach to extending the Dec-POMDP model with *macro-actions*, or temporally extended actions [Amato et al., 2014] does not assume such forms of communication. The formulation models prespecified behaviors as high-level actions (the macro-actions) in order to deal with significantly longer planning horizons. In particular, the approach extends the options framework Sutton et al. [1999] to Dec-POMDPs by using macro-actions, m_i, that execute a policy in a low-level Dec-POMDP until some terminal condition is met. We can then define policies over macro-actions for each agent, μ_i, for choosing macro-actions that depend on 'high-level observations' which are the termination conditions of the macro-actions (labeled with β in Figure 8.7).[4] Because macro-action policies are built from primitive actions, the value for high-level policies can be computed in a similar fashion, as described in Section 3.4.

As described in that section, the value can be expressed as an expectation

$$V(\mathbf{m}) = \mathbf{E}\left[\sum_{t=0}^{h-1} R(s_t, a_t) \,\middle|\, b_0, \mathbf{m}\right],$$

with the difference that the expectation now additionally is over macro-actions. In terms of computing the expectation, it is possible to define a recursive equation similar to (3.4.2) that explicitly deals with the cases that one or more macro-actions

---

[4] More generally, high-level observations can be defined that depend on these terminal conditions or underlying states.

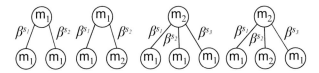

Fig. 8.7: A set of policies for one agent with macro-actions m_1 and m_1 and terminal conditions as β^s.

terminate. For details, we refer the reader toAmato et al. [2014]. The goal is to obtain a *hierarchically optimal joint macro-policy*. This is a joint macro-policy that produces the highest expected value that can be obtained by sequencing the agents' given macro-actions.

Two Dec-POMDP algorithms have been extended to consider macro-actions [Amato et al., 2014], but other extensions are possible. The key difference (as shown in Figure 8.7) is that nodes in a policy tree now select macro-actions (rather than primitive actions) and edges correspond to terminal conditions. Macro-action methods perform well in large domains when high-quality macro-actions are available. For example, consider the COOPERATIVE BOX PUSHING-inspired, multirobot domain shown in Figure 8.8. Here, robots were tasked with finding and retrieving boxes of two different sizes: large and small. The larger boxes can only be moved effectively by two robots and while the possible locations of boxes are known (in depots), the depot contents (the number and type of boxes) are unknown. Macro-actions were given that navigate to the depot locations, pick up the boxes, push them to the drop-off area and communicate any available messages.

The resulting behavior is illustrated in Figure (8.8), which shows screen captures from a video of the macro-policies in action. It is worth noting that the problem here represented as a flat Dec-POMDP is much larger than the COOPERATIVE BOX PUSHING benchmark problem (with over 10^9 states) and still the approach can generate high-quality solutions for a very long horizon [Amato et al., 2015b]. Additional approaches have extended these methods to controller-based solutions, to automatically generate macro-actions, and to remove the need for a full model of the underlying Dec-POMDP [Omidshafiei et al., 2015, Amato et al., 2015a].

8.3 Communication

Another more active direction of research involves incorporating communication between agents. We already saw in Section 2.4.3 that adding instantaneous, noise-free and cost-free communication allows us to transform a Dec-POMDP into an MPOMDP, but these assumptions are very strong. As such, research has tried identifying more reasonable assumptions and investigating how those impact the decision making processes.

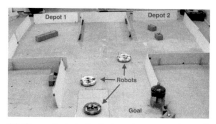

(a) One robot starts first and goes to depot 1 while the other robots begin moving towards the top middle.

(b) The robot in depot 1 sees a large box, so it turns on the red light (the light is not shown).

(c) The green robot sees light first, turns it off, and goes to depot 1. The white robot goes to depot 2.

(d) The robots in depot 1 push the large box and the robot in depot 2 pushes a small box to the goal.

Fig. 8.8: Multirobot video captures (signaling).

8.3.1 Implicit Communication and Explicit Communication

The main focus of this book is the regular Dec-POMDP, i.e., the setting without *explicitly modeled* communication. Nevertheless, in a Dec-POMDP the agents might very well communicate by means of their regular actions and regular observations. For instance, if one agent can use a chalk to write a mark on a blackboard and other agents have sensors to observe such a mark, the agents have a clear mechanism to send information from one to another, i.e., to communicate. We refer to such communication via regular ('domain') actions as implicit communication.[5] Furthermore, communication actions can be added to the action sets of each agent and communication observations can be added to the observation sets of each agent, allowing communication to be modeled in the same way as other actions and observations (e.g., with noise, delay or signal loss).

Definition 31 (Implicit and Explicit Communication). When a multiagent decision framework has a separate set of communication actions, we say that it supports *explicit communication*. Frameworks without explicit communication can and typically do still allow for *implicit communication*: the act of influencing the observations of one agent through the actions of another.

---

[5] Goldman and Zilberstein [2004] refer to this as 'indirect' communication.

We point out that the notion of implicit communication is very general. In fact, the only models that do not allow for implicit communication are precisely those that impose both transition and observation independence, such as the transition- and observation-independent Dec-MDP (Section 2.4.2) and the ND-POMDP (Section 8.1.1.2).

8.3.1.1 Explicit Communication Frameworks

When adding explicit communication to a multiagent framework, a first question to answer is what type of communication is supported. For instance, Xuan et al. [2001] identify three possible communication types:

- *Tell*, in which an agent informs another agent with some information (such as its local state, as assumed by Xuan et al.).
- *Query* in which an agent can ask another agent for some specific information.
- *Sync* in which, once initiated by at least one agent, all involved agents synchronize their knowledge (regarding an a priori specified set of things, such as the agents' local states).

Goldman and Zilberstein [2003] make a different categorization:

- *One-way communication,* in which information flow is unidirectional. This is similar to *tell*.
- *Two-way communication,* which leads to an exchange of messages. Both *Query* and *Sync* are forms of two-way communication. (Note that by issuing a query, the first agent may inform the second agent about aspects of the world about which it is uncertain.)
- *Acknowledged communication,* in which the agents send confirmation of the receipt of messages.

In addition, communication can differ as to which agents are involved: communication can be point-to-point, or broadcast to (a subset of) all agents.

There have been some approaches to extend the Dec-POMDP to explicitly incorporate communication actions and observations. For instance, the Dec-POMDP-Com[6] [Goldman and Zilberstein, 2003, 2004] additionally includes a set of messages and a cost function:

Definition 32 (Dec-POMDP-Com). A Dec-POMDP-Com is a tuple $\langle \mathcal{M}_{DecP}, \Sigma, C_\Sigma \rangle$, where

- $\mathcal{M}_{DecP}$ is a Dec-POMDP,
- Σ is the alphabet of possible messages that the agents can send,
- C_Σ is the communication cost function that indicates the cost of each possible message.

---

[6] The multiagent team decision problem (MTDP) that was mentioned in Section 2.2 has a similar extension, called the Com-MTDP [Pynadath and Tambe, 2002], which is equivalent to the Dec-POMDP-Com.

The messages in the Dec-POMDP-Com fulfill a threefold purpose: they serve as additional observations (i.e., they can be interpreted to serve as the set of auxiliary observations $\mathbb{Z}_i$, introduced in Section 2.4.4, for each agent i), they serve as additional actions (i.e., the mechanism by which the auxiliary observations are generated is by selecting these distinguished communication actions), and they serve as the basis for communication costs.

In the original paper by Goldman and Zilberstein [2003], the interaction proceeds as follows. At the beginning of every stage t, right after observing its individual observation $o_{i,t}$, each agent i gets the opportunity to first perform a communication action (i.e., broadcast one of the symbols from Σ). These messages are assumed to arrive instantly such that they are available to the other agents when deciding upon their next (domain) action a_i.

Of course, variations to this model are also possible, such as in Section 8.3.2, where we cover some approaches that deal with delayed communication. Also, the Dec-POMDP-Com model itself could allow different communication models, but most studies have considered noise-free instantaneous broadcast communication. That is, each agent broadcasts its message and receives the messages sent by all other agents instantaneously and without errors.

Although models with explicit communication seem more general than the models without, it is possible to transform the former to the latter. That is, a Dec-POMDP-Com can be transformed to a Dec-POMDP [Goldman and Zilberstein, 2004, Seuken and Zilberstein, 2008]. This means that contributions to the Dec-POMDP setting transfer to the case of general communication. Unfortunately, since the other way around, a Dec-POMDP is a special case of Dec-POMDP-Com, the computational hardness results also transfer to these models.

8.3.1.2 Updating of Information States and Semantics

It is important to realize that in a Dec-POMDP-Com the messages do not have any particular semantics. Instead, sending a particular message, say A, will lead (with some probability) to an observation for other agents, and now these other agents can condition their action on this observation. This means, that the goal for a Dec-POMDP-Com is to:

> find a joint policy that maximizes the expected total reward over the finite horizon. Solving for this policy embeds the *optimal meaning* of the messages chosen to be communicated.
> —Goldman and Zilberstein [2003]

That is, in this perspective the semantics of the communication actions become part of the optimization problem. This problem is considered by [Xuan et al., 2001, Goldman and Zilberstein, 2003, Spaan et al., 2006, Goldman et al., 2007, Amato et al., 2015b].

One can also consider the case where messages have *specified semantics*. In such a case the agents need a mechanism to process these semantics (i.e., to allow the messages to affect their internal state or beliefs). For instance, as we already discussed in Section 2.4.3, in an MPOMDP the agents share their local observations.

Each agent maintains a joint belief and performs an update of this joint belief, rather than maintaining the list of observations. In terms of the agent component of the multiagent decision process, introduced in Section 2.4.4, this means that the belief update function for each agent must be (at least partly) specified.[7]

It was shown by Pynadath and Tambe [2002] that for a Dec-POMDP-Com under instantaneous, noise-free and cost-free communication, a joint communication policy that shares the local observations at each stage (i.e., as in an MPOMDP) is optimal. Since this also makes intuitive sense, much research has investigated sharing local observations in models similar to the Dec-POMDP-Com [Pynadath and Tambe, 2002, Nair et al., 2004, Becker et al., 2005, Roth et al., 2005a,b, Spaan et al., 2006, Oliehoek et al., 2007, Roth et al., 2007, Goldman and Zilberstein, 2008]. The next two subsections cover observation-sharing approaches that try to lift some of the limiting assumptions: Section 8.3.2 allows for communication that is delayed one or more time steps and Section 8.3.3 deals with the case where broadcasting the local observation has nonzero costs.

8.3.2 Delayed Communication

Here we describe models, in which the agents can share their local observations via noise-free and cost-free communication, but where this communication can be delayed. That is, the assumption is that the synchronization of the agents such that each agent knows what the local observations of the other agents were takes one or more time steps.

8.3.2.1 One-Step Delayed Communication

We start investigating communication that arrives with a one-step delay (1-SD), which is referred to as the 'one-step delay observation sharing pattern' in the control literature [Witsenhausen, 1971, Varaiya and Walrand, 1978, Hsu and Marcus, 1982]. The consequence is that during execution at stage t the agents know $\bar{\theta}_{t-1}$, the joint action-observation history up to time step $t-1$, and the joint action a_{t-1} that was taken at the previous time step. Because all the agents know $\bar{\theta}_{t-1}$, they can compute the joint belief b_{t-1}, which is a Markovian signal. Therefore the agents do not need to maintain prior information; b_{t-1} takes the same role as the initial state distribution b_0 in a regular Dec-POMDP (i.e., without communication). Also, since we assume that during execution each agent knows the joint policy, each agent can infer the taken joint action a_{t-1}. However, the agents are uncertain regarding each other's last observation, and thus regarding the joint observation o_t. Effectively, this situation defines a collaborative Bayesian game (CBG) for each possible joint belief

---

[7] The Com-MTDP [Pynadath and Tambe, 2002] deals with this by introducing the notion of a 'richer' belief space and separating the belief update in a pre-communication and a post-communication part. The same functionality is also assumed in the Dec-POMDP-Com.

b_{t-1} and joint action a_{t-1} in which the type of each agent is its individual observation $o_{i,t}$. Note that these CBGs are different from the ones in Section (5.2.3); where the latter had types corresponding to the entire observation histories of agents, the CBGs considered here have types that correspond only to the last individual observation. Effectively, communication makes the earlier observations common information and therefore they need not be modeled in the type anymore.

As a result, the optimal value for the 1-SD setting can be expressed as:

$$V_1^{t,*}(b_t,a_t) = R(b_t,a_t) + \max_{\beta_{t+1}} \sum_{o_{t+1}\in\mathbb{O}} \Pr(o_{t+1}|b_t,a_t)V_1^{t+1,*}(b_{t+1},\beta_{t+1}(o_{t+1})), \quad (8.3.1)$$

where $\beta_{t+1} = \langle\beta_{1,t+1},\ldots,\beta_{n,t+1}\rangle$ is a joint BG policy containing individual BG policies $\beta_{i,t+1}$ that map individual observations to individual actions. For the motivation to indicate this value using the letter 'V' and for the relation to other value functions, see the discussion by Oliehoek [2010, Chap. 3]

Hsu and Marcus [1982] already showed that the value function for one-step delayed communication is piecewise linear and convex (PWLC), i.e., representable using sets of vectors. Not surprisingly, more recent approximation methods for POMDPs, e.g., Perseus [Spaan and Vlassis, 2005], can be transferred to its computation [Oliehoek et al., 2007]. Based on the insight that the value functions for both 0-step delayed communication (i.e., the MPOMDP) and 1-SD settings are representable using sets of vectors, Spaan et al. [2008] give a formulation that deals with stochastic delays of 0-1 time steps.

8.3.2.2 k-Steps Delayed Communication

We now consider the setting of k-step delayed communication[8] [Ooi and Wornell, 1996, Oliehoek, 2010, Nayyar et al., 2011, Oliehoek, 2013]. In this setting, at stage t each agent agent knows $\bar{\theta}_{t-k}$, the joint action-observation history of k stages earlier, and therefore can compute b_{t-k}, the joint belief induced by $\bar{\theta}_{t-k}$. Again, b_{t-k} is a Markov signal, so no further history needs to be retained and b_{t-k} takes the role of b_0 in the no-communication setting and b_{t-1} in the one-step delay setting. Indeed, one-step delay is just a special case of the k-step delay setting.

In contrast to the one-step delayed communication case, the agents do not know the last taken joint action. However, if we assume the agents know each other's policies, they do know q_{t-k}^k, the joint policy that has been executed during stages $t-k,\ldots,t-1$. This q_{t-k}^k is a length-k joint subtree policy rooted at stage $t-k$: it specifies a subtree policy $q_{i,t-k}^k$ for each agent i.

Let us assume that at a particular stage t the situation is as depicted in the top half of Figure 8.9: the system with two agents has $k=2$ steps delayed communication, so each agent knows b_{t-k} and q_{t-k}^k, the joint subtree policy that has been executed during stages $t-k,\ldots,t-1$. At this point, the agents need to select an

---

[8] This setting is referred to as the 'k-step delayed observation sharing information structure' in decentralized control theory.

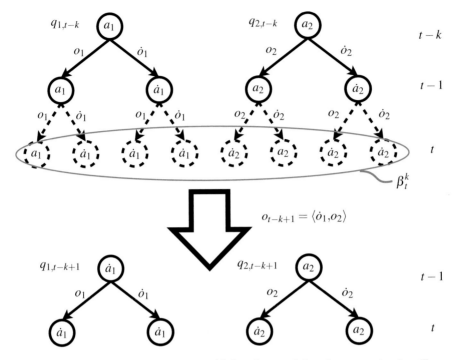

Fig. 8.9: Subtree policies in a system with $k = 2$ steps delayed communication. Top: policies at $t - k$. The policies are extended by a joint BG-policy β_t^k, shown dashed. Bottom: The resulting policies after joint observation $\langle \dot{o}_1, o_2 \rangle$.

action, but they do not know each other's individual observation history since stage $t - k$. That is, they have uncertainty with respect to the length-k observation history $\bar{o}_{t,|k|} = (o_{t-k+1}, \ldots, o_t)$. Effectively, this means that the agents have to use a joint BG-policy $\beta_t^k = \langle \beta_{1,t}^k, \ldots, \beta_{n,t}^k \rangle$ that implicitly maps length-k observation histories to joint actions $\beta_t^k(\bar{o}_{t,|k|}) = a_t$.

For example, let us assume that in the planning phase we computed a joint BG-policy β_t^k as indicated in the figure. As is shown, β_t^k can be used to extend the subtree policy q_{t-k}^k to form a longer subtree policy with $\tau = k + 1$ stage-to-go. Each agent has knowledge of this extended joint subtree policy $q_{t-k}^{k+1} = \langle q_{t-k}^k \circ \beta_t^k \rangle$. Consequently each agent i executes the action corresponding to its individual observation history $\beta_{i,t}^k(\bar{o}_{i,t,|k|}) = a_{i,t}$ and a transition occurs to stage $t + 1$. At that point each agent receives a new observation $o_{i,t+1}$ through perception and the joint observation o_{t-k+1} through communication, it transmits its individual observation, and it computes b_{t-k+1}. Now, all the agents know what action was taken at $t - k$ and what the following observation o_{t-k+1} was. Therefore the agents know which part of q_{t-k}^{k+1} has been executed during the last k stages $t - k + 1, \ldots, t$ and they discard the part not needed further; i.e., the joint observation 'consumes' part of the joint subtree policy: $q_{t-k+1}^k = q_{t-k}^{k+1} \Downarrow_{o_{t-k+1}}$ (see Definition 20).

Now the basic idea is that of Section 4.3.1: it is possible to define a plan-time MDP where the augmented states at stage t correspond to the common knowledge (i.e., (b_{t-k}, q_{t-k}^k)-pairs) and where actions correspond to joint BG-policies (β_t^k). Comparing to Section 4.3.1, the former correspond to φ_t and the latter to δ_t. Consequently, it is possible to define the value function of this plan-time MDP in a very similar way; see Oliehoek [2010] for details. Similarly to the development in Section 4.3.2, it turns out to be possible to replace the dependence on (b_{t-k}, q_{t-k}^k)-pairs by a plan-time sufficient statistic $\sigma_t(s_t, \bar{o}_{t,|k|})$ over states and joint length-k observation histories [Oliehoek, 2013], which in turn allows for a centralized formulation (a reduction to a POMDP), similar to the reformulation as an NOMDP of Section 4.5. The approach can be further generalized to exploit any common information that the agents might have [Nayyar et al., 2013, 2014].

8.3.3 Communication with Costs

Another way in which the strong assumptions of the MPOMDP can be relaxed is to assume that communication, while instantaneous and noise-free, no longer is cost-free. This can be a reasonable assumption in settings where the agents need to conserve energy (e.g., in robotic settings).

For instance, Becker et al. [2009] consider the question of *when* to communicate in a transition- and observation-independent Dec-MDP augmented with the 'sync' communication model. In more detail, each time step is separated in a domain action selected phase and a communication action selection phase. There are only two communication actions: *communicate* and *do not communicate*. When at least one agent chooses to communicate, synchronization is initiated and all agents participate in synchronizing their knowledge; in this case the agents suffer a particular communication cost.

Becker et al. [2009] investigate a myopic procedure for this setting: as long as no agent chooses to synchronize, all agents follow a decentralized (i.e, Dec-MDP) policy. At each stage, however, each agent estimates the *value of communication*—the difference between the expected value when communicating and when staying silent—by assuming that 1) in the future there will be no further possibility to communicate, and 2) other agents will not initiate communication. Since these assumptions introduce errors, Becker et al. also propose modified variants that mitigate these errors. In particular, the first assumption is modified by proposing a method to defer communicating if the value of communicating after one time step is higher than that of communicating now. The second assumption is overcome by modeling the myopic communication decision as a joint decision problem; essentially it is modeled as a collaborative Bayesian game in which the actions are *communicate* or *do not communicate,* while the types are the agents' local states.

8.3.4 Local Communication

The previous two subsections focused on softening the strong assumptions that the
MPOMDP model makes with respect to communication. However, even in settings
where instantaneous, noise-free and cost-free communication is available, broadcasting the individual observations to all other agents might not be feasible since
this scales poorly: with large numbers of agents the communication bandwidth may
become a problem. For instance, consider the setting of a Dec-MDP where each
agent i can observe a subset of state factors s_i (potentially overlapping with that of
other agents) that are the most relevant for its task. If all the agents can broadcast
their individual observation, each agent knows the complete state and the problem
reduces to that of an MMDP. When such broadcast communication is not possible,
however, the agents can still coordinate their actions using *local* coordination.

The main idea, introduced by Guestrin et al. [2002a] and Kok and Vlassis
[2006], is that we can approximate (without guarantees) the value function of an
MMDP using a factored value function (similar to the ND-POMDP discussed in
Section 8.1.1.2), which can be computed, for instance, via linear programming
[Guestrin et al., 2002a]. When we condition the resulting factored Q-function
$Q(s,a) \approx \sum_{i \in \mathbb{D}} Q^i(s_i, a_{\mathcal{N}(i)})$ on the state we implicitly define a coordination graph
$u(\cdot) \triangleq Q(s,\cdot)$, which allows the agents to coordinate their action selection online via
message passing (e.g., using max-sum or NDP) that only uses local communication.
The crucial insight that allows this to be applicable to Dec-MDPs is that in order to
condition the *local factor* Q^i on the current state s each agent only needs access to
its *local state* s_i: $u^i(\cdot) = Q^i(s_i,\cdot)$.

A somewhat related idea, introduced by Roth et al. [2007], is to minimize the
communication in a Dec-MDP by using the (exact or approximate) decision-tree
based solution of a factored MMDP. That is, certain solution methods for factored
MDPs (such as SPI [Boutilier et al., 2000], or SPUDD [Hoey et al., 1999]) produce
policies in the form of decision trees: the internal nodes specify state variables,
edges specify their values and the leaves specify the joint action to be taken for the
set of states corresponding to the path from root to leaf. Now the idea is that in some
cases (certain parts of) such a policy can be executed without communication even
if the agents observe different subsets of state variables. For instance, Figure 8.10a
shows the illustrative *relay world* in which two agents have a local state factor that
encodes their position. Each agent can perform 'shuffle' to randomly reset its location, 'exchange' a packet (only useful when both agents are at the top of the square)
or do nothing ('noop'). The optimal policy for agent 1 is shown in Figure 8.10b and
clearly demonstrates that requesting the location of the other agent via communication is only necessary when $\hat{s}_1 = L_1$. This idea has also been extended to partially
observable environments [Messias et al., 2011].

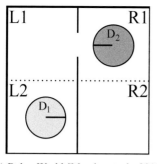
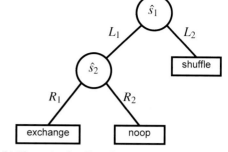

(a) Relay World [Messias et al., 2011].   (b) The optimal policy of agent 1 only depends on $\hat{s}_2$
(Reproduced with permission.)   when $\hat{s}_1 = L_1$.

Fig. 8.10: The structure of the policy for a factored multiagent problem can be exploited to reduce communication requirements.

8.4 Reinforcement Learning

This book focuses on planning for Dec-POMDPs, i.e., settings where the model of the environment is known at the start of the task. When this is not the case, we step into the realm of *multiagent reinforcement learning (MARL)*. In such settings, the model will have to be learned online (model-based MARL) or the agents will have to learn a solution directly without the use of a model (model-free methods). While there is a great deal of work on MARL in general [Panait and Luke, 2005, Buşoniu et al., 2008, Fudenberg and Levine, 2009, Tuyls and Weiss, 2012], MARL in partially observable settings has received little attention.

One of the main reasons for this gap in the literature seems to be that it is hard to properly define the setup of the reinforcement learning (RL) problem in these partially observable environments with multiple agents. For instance, it is not clear when or how the agents will the observe rewards.[9] Moreover, even when the agents can observe the state, general convergence of MARL under different assumptions is not fully understood: from the perspective of one agent, the environment has become nonstationary (since the other agent is also learning), which means that convergence guarantees from single-agent RL no longer hold. Claus and Boutilier [1998] argue that, in a cooperative setting, independent Q-learners are guaranteed to converge to a local optimum (but not necessarily to the global optimal solution). Nevertheless, this method has been reported to be successful in practice [e.g., Crites and Barto, 1998] and theoretical understanding of convergence of individual learners is progressing [e.g., Tuyls et al., 2006, Kaisers and Tuyls, 2010, Wunder et al., 2010]. There are *coupled* learning methods (e.g., Q-learning using the joint action space) that will converge to an optimal solution [Vlassis, 2007]. However, the guarantees

---

[9] Even in a single-agent POMDP, the agent is not assumed to have access to the immediate rewards, since they can convey hidden information about the states.

of these methods rely on the fact that the global states can be observed by all agents. In partially observable settings such guarantees have not yet been established. Nevertheless, a recent approach to *Bayesian RL* (i.e., the setting where there is a prior over models) in MPOMDPs demonstrates that learning in such settings is possible and scales to a moderate number of agents [Amato and Oliehoek, 2015].

Relatively few MARL approaches are applicable in partially observable settings where agents have only local observations. Peshkin et al. [2000] introduced *decentralized gradient ascent policy search (DGAPS)*, a method for MARL in partially observable settings based on gradient descent. DGAPS represents individual policies using finite-state controllers and assumes that agents observe the global rewards. Based on this, it is possible for each agent to independently update its policy in the direction of the gradient with respect to the return, resulting in a locally optimal joint policy. This approach was extended to learn policies for self-configurable modular robots [Varshavskaya et al., 2008]. Chang et al. [2004] also consider decentralized RL assuming that the global rewards are available to the agents. In their approach, these global rewards are interpreted as individual rewards, corrupted by noise due to the influence of other agents. Each agent explicitly tries to estimate the individual reward using Kalman filtering and performs independent Q-learning using the filtered individual rewards.

The methods by Wu et al. [2010b, 2013] are closely related to RL since they do not need entire models as input. They do, however, need access to a simulator which can be initialized to specific states. Similarly, Banerjee et al. [2012] iteratively learn policies for each agent using a sample-based version of the JESP algorithm where communication is used to alert the other agents that learning has been completed.

Finally, there are MARL methods for partially observed decentralized settings that require only limited amounts of communication. For instance, Boyan and Littman [1993] considered decentralized RL for a packet routing problem. Their approach, Q-routing, performs a type of Q-learning where there is only limited local communication: neighboring nodes communicate the expected future waiting time for a packet. Q-routing was extended to mobile wireless networks by Chang and Ho [2004]. A similar problem, distributed task allocation, is considered by Abdallah and Lesser [2007]. In this problem there also is a network, but now agents do not send communication packets, but rather tasks to neighbors. Again, communication is only local. This approach was extended to a hierarchical approach that includes so-called *supervisors* [Zhang et al., 2010]. The supervisors can communicate locally with other supervisors and with the agents they supervise ('workers'). Finally, in some RL methods for MMDPs (i.e., coupled methods) it is possible to have agents observe a subset of state factors if they have the ability to communicate *locally* [Guestrin et al., 2002b, Kok and Vlassis, 2006]. Such methods have been used in RoboCup soccer [Kok and Vlassis, 2005] and traffic control [Kuyer et al., 2008].

Chapter 9
Conclusion

This book gives an overview of the research performed since the early 2000s on decision making for multiagent systems under uncertainty. In particular, it focuses on the decentralized POMDP (Dec-POMDP) model, which is a general framework for modeling multiagent systems in settings that are both stochastic (i.e., the outcome of actions is uncertain) and partially observable (i.e., the state is uncertain). The core distinction between a Dec-POMDP and a (centralized) POMDP is that the execution phase is decentralized: each agent can only use its own observations to select its actions. This characteristic significantly changes the problem: there is no longer a compact sufficient statistic (or 'belief') that the agents can use to select actions, and the worst-case complexity of solving a Dec-POMDP is higher (NEXP-complete for the finite-horizon case). Such decentralized settings are important because they occur in many real-world applications, ranging from sensor networks to robotic teams. Moreover, in many of these settings dealing with uncertainty in a principled manner is important (e.g., avoiding critical failures while dealing with noisy sensors in robots problems or minimizing delays and thus economic cost due to traffic congestion while anticipating low-probability events that could lead to large disruptions). As such, Dec-POMDPs are a crucial framework for decision making in cooperative multiagent settings.

This book provides an overview of planning methods for both finite-horizon and infinite-horizon settings (which proceed for a finite or infinite number of time steps, respectively). Solution methods are provided that 1) are exact, 2) have some guarantees, or 3) are heuristic (have no guarantees but work well on larger benchmark domains). We also sketched some of the main lines of research that are currently being pursued: exploiting structure to increase scalability, employing hierarchical models, making more realistic assumptions with respect to communication, and dealing with settings where the model is not perfectly known in advance.

There are many big questions left to be answered in planning for Dec-POMDPs and we expect that much future research will continue to investigate these topics. In particular, the topics treated in Chapter 8 (exploiting structured models, hierarchical approaches, more versatile communication models, and reinforcement learning) all have seen quite significant advances in just the last few years. In parallel,

© The Author(s) 2016
F.A. Oliehoek and C. Amato, *A Concise Introduction to Decentralized POMDPs*,
SpringerBriefs in Intelligent Systems, DOI 10.1007/978-3-319-28929-8_9

due to some of the improvements in scalability, we see that the field is starting to shift from toy problems to benchmarks that, albeit still simplified, are motivated by real-world settings. Examples of such problems are settings for traffic control [Wu et al., 2013], communication network control [Winstein and Balakrishnan, 2013] and demonstrations on real multirobot systems [Emery-Montemerlo et al., 2005, Amato et al., 2015b]. We are hopeful that progress in these domains will inspire new ideas, and will attract the attention of both researchers and practitioners, thus leading to an application-driven influx of ideas to complement the traditionally theory-driven community studying these problems.

References

S. Abdallah and V. Lesser. Multiagent reinforcement learning and self-organization in a network of agents. In *Proceedings of the International Conference on Autonomous Agents and Multiagent Systems*, pages 172–179, 2007.

E. Altman. Applications of Markov decision processes in communication networks. In E. A. Feinberg and A. Shwartz, editors, *Handbook of Markov Decision Processes: Methods and Applications*, pages 489–536. Kluwer Academic Publishers, 2002.

C. Amato. *Increasing Scalability in Algorithms for Centralized and Decentralized Partially Observable Markov Decision Processes: Efficient Decision-Making and Coordination in Uncertain Environments*. PhD thesis, University of Massachusetts, Amherst, MA, 2010.

C. Amato. Cooperative decision making. In M. J. Kochenderfer, editor, *Decision Making Under Uncertainty: Theory and Application*, pages 159–182. MIT Press, 2015.

C. Amato and F. A. Oliehoek. Scalable planning and learning for multiagent POMDPs. In *Proceedings of the Twenty-Ninth AAAI Conference on Artificial Intelligence*, pages 1995–2002, 2015.

C. Amato and S. Zilberstein. Achieving goals in decentralized POMDPs. In *Proceedings of the International Conference on Autonomous Agents and Multiagent Systems*, pages 593–600, 2009.

C. Amato, D. S. Bernstein, and S. Zilberstein. Optimizing memory-bounded controllers for decentralized POMDPs. In *Proceedings of Uncertainty in Artificial Intelligence*, pages 1–8, 2007a.

C. Amato, A. Carlin, and S. Zilberstein. Bounded dynamic programming for decentralized POMDPs. In *Proceedings of the AAMAS Workshop on Multi-Agent Sequential Decision Making in Uncertain Domains (MSDM)*, 2007b.

C. Amato, J. S. Dibangoye, and S. Zilberstein. Incremental policy generation for finite-horizon DEC-POMDPs. In *Proceedings of the International Conference on Automated Planning and Scheduling*, pages 2–9, 2009.

© The Author(s) 2016
F.A. Oliehoek and C. Amato, *A Concise Introduction to Decentralized POMDPs*,
SpringerBriefs in Intelligent Systems, DOI 10.1007/978-3-319-28929-8

C. Amato, D. S. Bernstein, and S. Zilberstein. Optimizing fixed-size stochastic controllers for POMDPs and decentralized POMDPs. *Journal of Autonomous Agents and Multi-Agent Systems*, 21(3):293–320, 2010.

C. Amato, G. Chowdhary, A. Geramifard, N. K. Ure, and M. J. Kochenderfer. Decentralized control of partially observable Markov decision processes. In *Proceedings of the Fifty-Second IEEE Conference on Decision and Control*, pages 2398–2405, 2013.

C. Amato, G. Konidaris, and L. P. Kaelbling. Planning with macro-actions in decentralized POMDPs. In *Proceedings of the Thirteenth International Conference on Autonomous Agents and Multiagent Systems*, pages 1273–1280, 2014.

C. Amato, G. D. Konidaris, A. Anders, G. Cruz, J. P. How, and L. P. Kaelbling. Policy search for multi-robot coordination under uncertainty. In *Proceedings of the Robotics: Science and Systems Conference*, 2015a.

C. Amato, G. D. Konidaris, G. Cruz, C. A. Maynor, J. P. How, and L. P. Kaelbling. Planning for decentralized control of multiple robots under uncertainty. In *Proceedings of the International Conference on Robotics and Automation*, pages 1241–1248, 2015b.

T. Arai, E. Pagello, and L. Parker. Editorial: Advances in multirobot systems. *IEEE Transactions on Robotics and Automation*, 18(5):655–661, 2002.

R. Aras and A. Dutech. An investigation into mathematical programming for finite horizon decentralized POMDPs. *Journal of Artificial Intelligence Research*, 37: 329–396, 2010.

K. J. Åström. Optimal control of Markov processes with incomplete state information. *Journal of Mathematical Analysis and Applications*, 10(1):174–205, 1965.

R. I. Bahar, E. A. Frohm, C. M. Gaona, G. D. Hachtel, E. Macii, A. Pardo, and F. Somenzi. Algebraic decision diagrams and their applications. In *Proceedings of the 1993 IEEE/ACM International Conference on Computer-Aided Design*, pages 188–191, 1993.

H. Bai, D. Hsu, and W. S. Lee. Integrated perception and planning in the continuous space: A POMDP approach. *The International Journal of Robotics Research*, 33 (9):1288–1302, 2014.

B. Banerjee, J. Lyle, L. Kraemer, and R. Yellamraju. Sample bounded distributed reinforcement learning for decentralized POMDPs. In *Proceedings of the Twenty-Sixth AAAI Conference on Artificial Intelligence*, pages 1256–1262, 2012.

A. L. C. Bazzan. A distributed approach for coordination of traffic signal agents. *Journal of Autonomous Agents and Multi-Agent Systems*, 10(1):131–164, 2005.

A. L. C. Bazzan, D. de Oliveira, and B. C. da Silva. Learning in groups of traffic signals. *Engineering Applications of Artificial Intelligence*, 23(4):560–568, 2010.

R. Becker, S. Zilberstein, V. Lesser, and C. V. Goldman. Transition-independent decentralized Markov decision processes. In *Proceedings of the International Conference on Autonomous Agents and Multiagent Systems*, pages 41–48, 2003.

R. Becker, S. Zilberstein, and V. Lesser. Decentralized Markov decision processes with event-driven interactions. In *Proceedings of the International Conference on Autonomous Agents and Multiagent Systems*, pages 302–309, 2004a.

R. Becker, S. Zilberstein, V. Lesser, and C. V. Goldman. Solving transition independent decentralized Markov decision processes. *Journal of Artificial Intelligence Research*, 22:423–455, 2004b.

R. Becker, V. Lesser, and S. Zilberstein. Analyzing myopic approaches for multi-agent communication. In *Proceedings of the International Conference on Intelligent Agent Technology*, pages 550–557, 2005.

R. Becker, A. Carlin, V. Lesser, and S. Zilberstein. Analyzing myopic approaches for multi-agent communication. *Computational Intelligence*, 25(1):31–50, 2009.

D. Bellhouse. The problem of Waldegrave. *Journal Électronique d'Histoire des Probabilités et de la Statistique*, 3(2):1–12, 2007.

R. Bellman. *Dynamic Programming*. Princeton University Press, 1957.

D. S. Bernstein, R. Givan, N. Immerman, and S. Zilberstein. The complexity of decentralized control of Markov decision processes. *Mathematics of Operations Research*, 27(4):819–840, 2002.

D. S. Bernstein, E. A. Hansen, and S. Zilberstein. Bounded policy iteration for decentralized POMDPs. In *Proceedings of the International Joint Conference on Artificial Intelligence*, pages 1287–1292, 2005.

D. S. Bernstein, C. Amato, E. A. Hansen, and S. Zilberstein. Policy iteration for decentralized control of Markov decision processes. *Journal of Artificial Intelligence Research*, 34:89–132, 2009.

U. Bertele and F. Brioschi. *Nonserial Dynamic Programming*. Academic Press, 1972.

D. P. Bertsekas. *Dynamic Programming and Optimal Control*, volume II. Athena Scientific, 3rd edition, 2007.

K. Binmore. *Fun and Games*. D.C. Heath and Company, 1992.

C. M. Bishop. *Pattern Recognition and Machine Learning*. Springer, 2006.

R. Bordini, M. Dastani, J. Dix, and A. El Fallah Segrouchni, editors. *Multi-Agent Programming: Languages, Platforms and Applications*. Springer, 2005.

A. Boularias and B. Chaib-draa. Exact dynamic programming for decentralized POMDPs with lossless policy compression. In *Proceedings of the International Conference on Automated Planning and Scheduling*, pages 20–28, 2008.

C. Boutilier. Planning, learning and coordination in multiagent decision processes. In *Proceedings of the 6th Conference on Theoretical Aspects of Rationality and Knowledge*, pages 195–210, 1996.

C. Boutilier, T. Dean, and S. Hanks. Decision-theoretic planning: Structural assumptions and computational leverage. *Journal of Artificial Intelligence Research*, 11: 1–94, 1999.

C. Boutilier, R. Dearden, and M. Goldszmidt. Stochastic dynamic programming with factored representations. *Artificial Intelligence*, 121(1-2):49–107, 2000.

J. A. Boyan and M. L. Littman. Packet routing in dynamically changing networks: A reinforcement learning approach. In *Advances in Neural Information Processing Systems 6*, pages 671–678, 1993.

M. E. Bratman. *Intention, Plans and Practical Reason*. Harvard University Press, 1987.

L. Buşoniu, R. Babuška, and B. De Schutter. A comprehensive survey of multi-agent reinforcement learning. *IEEE Transactions on Systems, Man, and Cybernetics, Part C: Applications and Reviews*, 38(2):156–172, 2008.

A. Carlin and S. Zilberstein. Value-based observation compression for DEC-POMDPs. In *Proceedings of the International Conference on Autonomous Agents and Multiagent Systems*, pages 501–508, 2008.

A. R. Cassandra. *Exact and approximate algorithms for partially observable Markov decision processes*. PhD thesis, Brown University, 1998.

Y.-H. Chang and T. Ho. Mobilized ad-hoc networks: A reinforcement learning approach. In *Proceedings of the First International Conference on Autonomic Computing*, pages 240–247, 2004.

Y.-H. Chang, T. Ho, and L. P. Kaelbling. All learning is local: Multi-agent learning in global reward games. In *Advances in Neural Information Processing Systems 16*, pages 807–814, 2004.

H.-T. Cheng. *Algorithms for partially observable Markov decision processes*. PhD thesis, University of British Columbia, 1988.

C. Claus and C. Boutilier. The dynamics of reinforcement learning in cooperative multiagent systems. In *Proceedings of the National Conference on Artificial Intelligence*, pages 746–752, 1998.

R. Cogill, M. Rotkowitz, B. V. Roy, and S. Lall. An approximate dynamic programming approach to decentralized control of stochastic systems. In *Proceedings of the Forty-Second Allerton Conference on Communication, Control, and Computing*, pages 1040–1049, 2004.

P. R. Cohen and H. J. Levesque. Intention is choice with commitment. *Artificial Intelligence*, 42(3):213–261, 1990.

P. R. Cohen and H. J. Levesque. Confirmations and joint action. In *Proceedings of the International Joint Conference on Artificial Intelligence*, pages 951–957. Morgan Kaufmann, 1991a.

P. R. Cohen and H. J. Levesque. Teamwork. *Nous*, 25(4):487–512, 1991b.

R. H. Crites and A. G. Barto. Elevator group control using multiple reinforcement learning agents. *Machine Learning*, 33(2-3):235–262, 1998.

R. Davis and R. G. Smith. Negotiation as a metaphor for distributed problem solving. *Artificial Intelligence*, 20(1):63–109, 1983.

R. Dechter. Bucket elimination: A unifying framework for processing hard and soft constraints. *Constraints*, 2(1):51–55, 1997.

R. Dechter. Reasoning with probabilistic and deterministic graphical models: Exact algorithms. *Synthesis Lectures on Artificial Intelligence and Machine Learning*, 7(3):1–191, 2013.

M. desJardins, E. H. Durfee, C. L. Ortiz Jr., and M. Wolverton. A survey of research in distributed, continual planning. *AI Magazine*, 20(4):13–22, 1999.

J. S. Dibangoye, A.-I. Mouaddib, and B. Chai-draa. Point-based incremental pruning heuristic for solving finite-horizon DEC-POMDPs. In *Proceedings of the International Conference on Autonomous Agents and Multiagent Systems*, pages 569–576, 2009.

J. S. Dibangoye, C. Amato, O. Buffet, and F. Charpillet. Optimally solving Dec-POMDPs as continuous-state MDPs. In *Proceedings of the International Joint Conference on Artificial Intelligence*, pages 90–96, 2013.

J. S. Dibangoye, C. Amato, O. Buffet, and F. Charpillet. Exploiting separability in multiagent planning with continuous-state MDPs. In *Proceedings of the Thirteenth International Conference on Autonomous Agents and Multiagent Systems*, pages 1281–1288, 2014.

E. H. Durfee. Distributed problem solving and planning. In *Multi-agents systems and applications*, pages 118–149. Springer, 2001.

E. H. Durfee, V. R. Lesser, and D. D. Corkill. Coherent cooperation among communicating problem solvers. *Computers, IEEE Transactions on*, 100(11):1275–1291, 1987.

B. Eker and H. L. Akın. Using evolution strategies to solve DEC-POMDP problems. *Soft Computing—A Fusion of Foundations, Methodologies and Applications*, 14 (1):35–47, 2010.

R. Emery-Montemerlo. *Game-Theoretic Control for Robot Teams*. PhD thesis, Carnegie Mellon University, 2005.

R. Emery-Montemerlo, G. Gordon, J. Schneider, and S. Thrun. Approximate solutions for partially observable stochastic games with common payoffs. In *Proceedings of the International Conference on Autonomous Agents and Multiagent Systems*, pages 136–143, 2004.

R. Emery-Montemerlo, G. Gordon, J. Schneider, and S. Thrun. Game theoretic control for robot teams. In *Proceedings of the International Conference on Robotics and Automation*, pages 1175–1181, 2005.

A. Farinelli, A. Rogers, A. Petcu, and N. R. Jennings. Decentralised coordination of low-power embedded devices using the max-sum algorithm. In *Proceedings of the International Conference on Autonomous Agents and Multiagent Systems*, pages 639–646, 2008.

Z. Feng and E. Hansen. Approximate planning for factored POMDPs. In *Proceedings of the Sixth European Conference on Planning*, pages 99–106, 2001.

D. Fudenberg and D. K. Levine. Learning and equilibrium. *Annual Review of Economics*, 1:385–420, 2009.

A.-J. Garcia-Sanchez, F. Garcia-Sanchez, F. Losilla, P. Kulakowski, J. Garcia-Haro, A. Rodríguez, J.-V. López-Bao, and F. Palomares. Wireless sensor network deployment for monitoring wildlife passages. *Sensors*, 10(8):7236–7262, 2010.

M. P. Georgeff, B. Pell, M. E. Pollack, M. Tambe, and M. Wooldridge. The belief-desire-intention model of agency. In *Proceedings of the 5th International Workshop on Intelligent Agents V, Agent Theories, Architectures, and Languages*, pages 1–10, 1999.

P. J. Gmytrasiewicz and P. Doshi. A framework for sequential planning in multiagent settings. *Journal of Artificial Intelligence Research*, 24:49–79, 2005.

P. J. Gmytrasiewicz and E. H. Durfee. A rigorous, operational formalization of recursive modeling. In *Proceedings of the First International Conference on Multiagent Systems*, pages 125–132, 1995.

P. J. Gmytrasiewicz, S. Noh, and T. Kellogg. Bayesian update of recursive agent models. *User Modeling and User-Adapted Interaction*, 8(1-2):49–69, 1998.

C. V. Goldman and S. Zilberstein. Optimizing information exchange in cooperative multi-agent systems. In *Proceedings of the International Conference on Autonomous Agents and Multiagent Systems*, pages 137–144, 2003.

C. V. Goldman and S. Zilberstein. Decentralized control of cooperative systems: Categorization and complexity analysis. *Journal of Artificial Intelligence Research*, 22:143–174, 2004.

C. V. Goldman and S. Zilberstein. Communication-based decomposition mechanisms for decentralized MDPs. *Journal of Artificial Intelligence Research*, 32: 169–202, 2008.

C. V. Goldman, M. Allen, and S. Zilberstein. Learning to communicate in a decentralized environment. *Journal of Autonomous Agents and Multi-Agent Systems*, 15(1):47–90, 2007.

B. J. Grosz and S. Kraus. Collaborative plans for complex group action. *Artificial Intelligence*, 86(2):269–357, 1996.

B. J. Grosz and C. Sidner. Plans for discourse. In *Intentions in Communication*, pages 417–444. MIT Press, 1990.

B. J. Grosz and C. L. Sidner. Attention, intentions, and the structure of discourse. *Computational linguistics*, 12(3):175–204, 1986.

M. Grześ, P. Poupart, and J. Hoey. Isomorph-free branch and bound search for finite state controllers. In *Proceedings of the International Joint Conference on Artificial Intelligence*, pages 2282–2290. AAAI Press, 2013.

C. Guestrin, D. Koller, and R. Parr. Multiagent planning with factored MDPs. In *Advances in Neural Information Processing Systems 14*, pages 1523–1530, 2002a.

C. Guestrin, M. Lagoudakis, and R. Parr. Coordinated reinforcement learning. In *Proceedings of the International Conference on Machine Learning*, pages 227–234, 2002b.

J. Y. Halpern. *Reasoning about Uncertainty*. MIT Press, 2003.

E. A. Hansen. Solving POMDPs by searching in policy space. In *Proceedings of the Fourteenth Conference on Uncertainty in Artificial Intelligence*, pages 211–219, 1998.

E. A. Hansen, D. S. Bernstein, and S. Zilberstein. Dynamic programming for partially observable stochastic games. In *Proceedings of the National Conference on Artificial Intelligence*, pages 709–715, 2004.

Y.-C. Ho. Team decision theory and information structures. *Proceedings of the IEEE*, 68(6):644–654, 1980.

J. Hoey, R. St-Aubin, A. J. Hu, and C. Boutilier. SPUDD: Stochastic planning using decision diagrams. In *Proceedings of Uncertainty in Artificial Intelligence*, pages 279–288, 1999.

J. E. Hopcroft and J. D. Ullman. *Introduction to Automata Theory, Languages, and Computation*. Addison-Wesley, 1979.

R. A. Howard and J. E. Matheson. Influence diagrams. In *The Principles and Applications of Decision Analysis, Vol. II.*, pages 719–763. Strategic Decisions Group, 1984.

K. Hsu and S. Marcus. Decentralized control of finite state Markov processes. *IEEE Transactions on Automatic Control*, 27(2):426–431, 1982.

M. N. Huhns, editor. *Distributed Artificial Intelligence*. Pitman Publishing, 1987.

N. R. Jennings. Controlling cooperative problem solving in industrial multi-agent systems using joint intentions. *Artificial Intelligence*, 75(2):195–240, 1995.

N. R. Jennings. Agent-based computing: Promise and perils. In *Proceedings of the Sixteenth International Joint Conference on Artificial Intelligence*, pages 1429–1436, 1999.

L. P. Kaelbling and T. Lozano-Pérez. Integrated task and motion planning in belief space. *The International Journal of Robotics Research*, 32(9-10):1194–1227, 2013.

L. P. Kaelbling, M. Littman, and A. Moore. Reinforcement learning: A survey. *Journal of Artificial Intelligence Research*, 4:237–285, 1996.

L. P. Kaelbling, M. L. Littman, and A. R. Cassandra. Planning and acting in partially observable stochastic domains. *Artificial Intelligence*, 101(1-2):99–134, 1998.

M. Kaisers and K. Tuyls. Frequency adjusted multi-agent Q-learning. In *Proceedings of the International Conference on Autonomous Agents and Multiagent Systems*, pages 309–316, 2010.

R. E. Kalman. A new approach to linear filtering and prediction problems. *Transactions of the ASME–Journal of Basic Engineering*, 82(Series D):35–45, 1960.

M. J. Kearns. Graphical games. In N. Nisan, T. Roughgarden, E. Tardos, and V. Vazirani, editors, *Algorithmic Game Theory*. Cambridge University Press, 2007.

M. J. Kearns, M. L. Littman, and S. P. Singh. Graphical models for game theory. In *Proceedings of Uncertainty in Artificial Intelligence*, pages 253–260, 2001.

K. K. Khedo, R. Perseedoss, and A. Mungur. A wireless sensor network air pollution monitoring system. *International Journal of Wireless & Mobile Networks*, 2(2):31–45, 2010.

Y. Kim, R. Nair, P. Varakantham, M. Tambe, and M. Yokoo. Exploiting locality of interaction in networked distributed POMDPs. In *Proceedings of the AAAI Spring Symposium on Distributed Plan and Schedule Management*, pages 41–48, 2006.

D. Kinny and M. Georgeff. Modelling and design of multi-agent systems. In *Intelligent Agents III Agent Theories, Architectures, and Languages*, pages 1–20. Springer, 1997.

H. Kitano, S. Tadokoro, I. Noda, H. Matsubara, T. Takahashi, A. Shinjoh, and S. Shimada. RoboCup Rescue: Search and rescue in large-scale disasters as a domain for autonomous agents research. In *Proceedings of the International Conference on Systems, Man and Cybernetics*, pages 739–743, 1999.

M. J. Kochenderfer, C. Amato, G. Chowdhary, J. P. How, H. J. D. Reynolds, J. R. Thornton, P. A. Torres-Carrasquillo, N. K. Üre, and J. Vian. *Decision making under uncertainty: theory and application*. MIT Press, 2015.

J. R. Kok and N. Vlassis. Using the max-plus algorithm for multiagent decision making in coordination graphs. In *RoboCup-2005: Robot Soccer World Cup IX*, pages 1–12, 2005.

J. R. Kok and N. Vlassis. Collaborative multiagent reinforcement learning by payoff propagation. *Journal of Machine Learning Research*, 7:1789–1828, 2006.

D. Koller and N. Friedman. *Probabilistic Graphical Models: Principles and Techniques*. MIT Press, 2009.

D. Koller and A. Pfeffer. Representations and solutions for game-theoretic problems. *Artificial Intelligence*, 94(1-2):167–215, 1997.

A. Kumar and S. Zilberstein. Constraint-based dynamic programming for decentralized POMDPs with structured interactions. In *Proceedings of the International Conference on Autonomous Agents and Multiagent Systems*, pages 561–568, 2009.

A. Kumar and S. Zilberstein. Point-based backup for decentralized POMDPs: Complexity and new algorithms. In *Proceedings of the International Conference on Autonomous Agents and Multiagent Systems*, pages 1315–1322, 2010a.

A. Kumar and S. Zilberstein. Anytime planning for decentralized POMDPs using expectation maximization. In *Proceedings of Uncertainty in Artificial Intelligence*, pages 294–301, 2010b.

A. Kumar, S. Zilberstein, and M. Toussaint. Scalable multiagent planning using probabilistic inference. In *Proceedings of the International Joint Conference on Artificial Intelligence*, pages 2140–2146, 2011.

L. Kuyer, S. Whiteson, B. Bakker, and N. Vlassis. Multiagent reinforcement learning for urban traffic control using coordination graphs. In *Machine Learning and Knowledge Discovery in Databases*, Lecture Notes in Computer Science, pages 656–671, 2008.

V. Lesser, C. L. Ortiz Jr., and M. Tambe, editors. *Distributed Sensor Networks: A Multiagent Perspective*, volume 9. Kluwer Academic Publishers, 2003.

V. R. Lesser. Cooperative multiagent systems: A personal view of the state of the art. *IEEE Transactions on Knowledge and Data Engineering*, 11:133–142, 1999.

K. Leyton-Brown and Y. Shoham. *Essentials of Game Theory: A Concise Multidisciplinary Introduction*. Synthesis Lectures on Artificial Intelligence and Machine Learning. Morgan & Claypool, 2008.

M. Littman, A. Cassandra, and L. Kaelbling. Learning policies for partially observable environments: Scaling up. In *Proceedings of the International Conference on Machine Learning*, pages 362–370, 1995.

J. Liu and K. P. Sycara. Exploiting problem structure for distributed constraint optimization. In *Proceedings of the International Conference on Multiagent Systems*, pages 246–253, 1995.

L. C. MacDermed and C. Isbell. Point based value iteration with optimal belief compression for Dec-POMDPs. In *Advances in Neural Information Processing Systems 26*, pages 100–108, 2013.

O. Madani, S. Hanks, and A. Condon. On the undecidability of probabilistic planning and infinite-horizon partially observable Markov decision problems. In *Proceedings of the National Conference on Artificial Intelligence*, pages 541–548, 1999.

A. Mahajan and M. Mannan. Decentralized stochastic control. *Annals of Operations Research*, pages 1–18, 2014.

S. Mannor, R. Rubinstein, and Y. Gat. The cross entropy method for fast policy search. In *Proceedings of the International Conference on Machine Learning*, pages 512–519, 2003.

J. Marecki, T. Gupta, P. Varakantham, M. Tambe, and M. Yokoo. Not all agents are equal: Scaling up distributed POMDPs for agent networks. In *Proceedings of the International Conference on Autonomous Agents and Multiagent Systems*, pages 485–492, 2008.

R. Marinescu and R. Dechter. And/or branch-and-bound search for combinatorial optimization in graphical models. *Artificial Intelligence*, 173(16-17):1457–1491, 2009.

J. Marschak. Elements for a theory of teams. *Management Science*, 1:127–137, 1955.

J. Marschak and R. Radner. *Economic Theory of Teams*. Yale University Press, 1972.

J. V. Messias. *Decision-Making under Uncertainty for Real Robot Teams*. PhD thesis, Institute for Systems and Robotics, Instituto Superior Técnico, 2014.

J. V. Messias, M. Spaan, and P. U. Lima. Efficient offline communication policies for factored multiagent POMDPs. In *Advances in Neural Information Processing Systems 24*, pages 1917–1925, 2011.

N. Meuleau, K. Kim, L. P. Kaelbling, and A. R. Cassandra. Solving POMDPs by searching the space of finite policies. In *Proceedings of the Fifteenth Conference on Uncertainty in Artificial Intelligence*, pages 417–426, 1999a.

N. Meuleau, L. Peshkin, K.-E. Kim, and L. P. Kaelbling. Learning finite-state controllers for partially observable environments. In *Proceedings of the Fifteenth Conference on Uncertainty in Artificial Intelligence*, pages 427–436, 1999b.

P. J. Modi, W.-M. Shen, M. Tambe, and M. Yokoo. Adopt: Asynchronous distributed constraint optimization with quality guarantees. *Artificial Intelligence*, 161:149–180, 2005.

H. Mostafa and V. Lesser. Offline planning for communication by exploiting structured interactions in decentralized MDPs. In *Proceedings of 2009 IEEE/WIC/ACM International Conference on Web Intelligence and Intelligent Agent Technology*, pages 193–200, 2009.

H. Mostafa and V. Lesser. A compact mathematical formulation for problems with structured agent interactions. In *Proceedings of the AAMAS Workshop on Multi-Agent Sequential Decision Making in Uncertain Domains (MSDM)*, pages 55–62, 2011a.

H. Mostafa and V. R. Lesser. Compact mathematical programs for DEC-MDPs with structured agent interactions. In *Proceedings of the Twenty-Seventh Conference on Uncertainty in Artificial Intelligence*, pages 523–530, 2011b.

R. Nair and M. Tambe. Hybrid BDI-POMDP framework for multiagent teaming. *Journal of Artificial Intelligence Research*, 23:367–420, 2005.

R. Nair, M. Tambe, and S. Marsella. Team formation for reformation. In *Proceedings of the AAAI Spring Symposium on Intelligent Distributed and Embedded Systems*, pages 52–56, 2002.

R. Nair, M. Tambe, and S. Marsella. Role allocation and reallocation in multiagent teams: Towards a practical analysis. In *Proceedings of the International Conference on Autonomous Agents and Multiagent Systems*, pages 552–559, 2003a.

R. Nair, M. Tambe, and S. Marsella. Team formation for reformation in multiagent domains like RoboCupRescue. In *Proceedings of RoboCup-2002 International Symposium*, Lecture Notes in Computer Science, pages 150–161, 2003b.

R. Nair, M. Tambe, M. Yokoo, D. V. Pynadath, and S. Marsella. Taming decentralized POMDPs: Towards efficient policy computation for multiagent settings. In *Proceedings of the International Joint Conference on Artificial Intelligence*, pages 705–711, 2003c.

R. Nair, M. Roth, and M. Yokoo. Communication for improving policy computation in distributed POMDPs. In *Proceedings of the International Conference on Autonomous Agents and Multiagent Systems*, pages 1098–1105, 2004.

R. Nair, P. Varakantham, M. Tambe, and M. Yokoo. Networked distributed POMDPs: A synthesis of distributed constraint optimization and POMDPs. In *Proceedings of the National Conference on Artificial Intelligence*, pages 133–139, 2005.

A. Nayyar, A. Mahajan, and D. Teneketzis. Optimal control strategies in delayed sharing information structures. *IEEE Transactions on Automatic Control*, 56(7): 1606–1620, 2011.

A. Nayyar, A. Mahajan, and D. Teneketzis. Decentralized stochastic control with partial history sharing: A common information approach. *IEEE Transactions on Automatic Control*, 58:1644–1658, 2013.

A. Nayyar, A. Mahajan, and D. Teneketzis. The common-information approach to decentralized stochastic control. In G. Como, B. Bernhardsson, and A. Rantzer, editors, *Information and Control in Networks*, volume 450 of *Lecture Notes in Control and Information Sciences*, pages 123–156. Springer, 2014.

R. T. Van Katwijk. *Multi-Agent Look-Ahead Traffic-Adaptive Control*. PhD thesis, Delft University of Technology, 2008.

F. A. Oliehoek. *Value-Based Planning for Teams of Agents in Stochastic Partially Observable Environments*. PhD thesis, Informatics Institute, University of Amsterdam, 2010.

F. A. Oliehoek. Decentralized POMDPs. In M. Wiering and M. van Otterlo, editors, *Reinforcement Learning: State of the Art*, volume 12 of *Adaptation, Learning, and Optimization*, pages 471–503. Springer, 2012.

F. A. Oliehoek. Sufficient plan-time statistics for decentralized POMDPs. In *Proceedings of the Twenty-Third International Joint Conference on Artificial Intelligence*, pages 302–308, 2013.

F. A. Oliehoek and C. Amato. Dec-POMDPs as non-observable MDPs. IAS technical report IAS-UVA-14-01, Intelligent Systems Lab, University of Amsterdam, Amsterdam, The Netherlands, 2014.

F. A. Oliehoek and A. Visser. A hierarchical model for decentralized fighting of large scale urban fires. In *AAMAS Workshop on Hierarchical Autonomous Agents and Multi-Agent Systems*, pages 14–21, 2006.

F. A. Oliehoek, M. T. J. Spaan, and N. Vlassis. Dec-POMDPs with delayed communication. In *Proceedings of the AAMAS Workshop on Multi-Agent Sequential Decision Making in Uncertain Domains (MSDM)*, 2007.

F. A. Oliehoek, J. F. Kooi, and N. Vlassis. The cross-entropy method for policy search in decentralized POMDPs. *Informatica*, 32:341–357, 2008a.

F. A. Oliehoek, M. T. J. Spaan, and N. Vlassis. Optimal and approximate Q-value functions for decentralized POMDPs. *Journal of Artificial Intelligence Research*, 32:289–353, 2008b.

F. A. Oliehoek, M. T. J. Spaan, S. Whiteson, and N. Vlassis. Exploiting locality of interaction in factored Dec-POMDPs. In *Proceedings of the Seventh Joint International Conference on Autonomous Agents and Multiagent Systems*, pages 517–524, 2008c.

F. A. Oliehoek, S. Whiteson, and M. T. J. Spaan. Lossless clustering of histories in decentralized POMDPs. In *Proceedings of the Eighth International Conference on Autonomous Agents and Multiagent Systems*, pages 577–584, 2009.

F. A. Oliehoek, M. T. J. Spaan, J. Dibangoye, and C. Amato. Heuristic search for identical payoff Bayesian games. In *Proceedings of the Ninth International Conference on Autonomous Agents and Multiagent Systems*, pages 1115–1122, 2010.

F. A. Oliehoek, S. Whiteson, and M. T. J. Spaan. Exploiting structure in cooperative Bayesian games. In *Proceedings of the Twenty-Eighth Conference on Uncertainty in Artificial Intelligence*, pages 654–664, 2012a.

F. A. Oliehoek, S. Witwicki, and L. P. Kaelbling. Influence-based abstraction for multiagent systems. In *Proceedings of the Twenty-Sixth AAAI Conference on Artificial Intelligence*, pages 1422–1428, 2012b.

F. A. Oliehoek, M. T. J. Spaan, C. Amato, and S. Whiteson. Incremental clustering and expansion for faster optimal planning in decentralized POMDPs. *Journal of Artificial Intelligence Research*, 46:449–509, 2013a.

F. A. Oliehoek, S. Whiteson, and M. T. J. Spaan. Approximate solutions for factored Dec-POMDPs with many agents. In *Proceedings of the Twelfth International Conference on Autonomous Agents and Multiagent Systems*, pages 563–570, 2013b.

F. A. Oliehoek, M. T. J. Spaan, and S. Witwicki. Factored upper bounds for multiagent planning problems under uncertainty with non-factored value functions. In *Proceedings of the Twenty-Fourth International Joint Conference on Artificial Intelligence*, pages 1645–1651, 2015a.

F. A. Oliehoek, M. T. J. Spaan, and S. Witwicki. Influence-optimistic local values for multiagent planning—extended version. *ArXiv e-prints*, arXiv:1502.05443, 2015b.

S. Omidshafiei, A. Agha-mohammadi, C. Amato, and J. P. How. Decentralized control of partially observable Markov decision processes using belief space macro-actions. In *Proceedings of the International Conference on Robotics and Automation*, pages 5962–5969, 2015.

S. Omidshafiei, A. Agha-mohammadi, C. Amato, S.-Y. Liu, J. P. How, and J. Vian. Graph-based cross entropy method for solving multi-robot decentralized POMDPs. In *Proceedings of the International Conference on Robotics and Automation*, 2016.

J. M. Ooi and G. W. Wornell. Decentralized control of a multiple access broadcast channel: Performance bounds. In *Proceedings of the 35th Conference on Decision and Control*, pages 293–298, 1996.

M. J. Osborne and A. Rubinstein. *A Course in Game Theory*. The MIT Press, 1994.

Y. Ouyang and D. Teneketzis. Balancing through signaling in decentralized routing. In *Proceedings of the 53rd Conference on Decision and Control*, pages 1675–1680, 2014.

J. Pajarinen and J. Peltonen. Efficient planning for factored infinite-horizon DEC-POMDPs. In *Proceedings of the International Joint Conference on Artificial Intelligence*, pages 325–331, 2011a.

J. Pajarinen and J. Peltonen. Expectation maximization for average reward decentralized POMDPs. In *Machine Learning and Knowledge Discovery in Databases*, pages 129–144. Springer, 2013.

J. K. Pajarinen and J. Peltonen. Periodic finite state controllers for efficient POMDP and DEC-POMDP planning. In *Advances in Neural Information Processing Systems 24*, pages 2636–2644, 2011b.

L. Panait and S. Luke. Cooperative multi-agent learning: The state of the art. *Journal of Autonomous Agents and Multi-Agent Systems*, 11(3):387–434, 2005.

C. H. Papadimitriou and J. N. Tsitsiklis. The complexity of Markov decision processes. *Mathematics of Operations Research*, 12(3):441–451, 1987.

S. Paquet, L. Tobin, and B. Chaib-draa. An online POMDP algorithm for complex multiagent environments. In *Proceedings of the International Conference on Autonomous Agents and Multiagent Systems*, pages 970–977, 2005.

G. Pavlin, F. C. A. Groen, P. de Oude, and M. Kamermans. A distributed approach to gas detection and source localization using heterogeneous information. In R. Babuska and F. C. A. Groen, editors, *Interactive Collaborative Information Systems*, volume 281 of *Studies in Computational Intelligence*, pages 453–474. Springer, 2010.

J. Pearl. *Probabilistic Reasoning In Intelligent Systems: Networks of Plausible Inference*. Morgan Kaufmann, 1988.

L. Peshkin. *Reinforcement Learning by Policy Search*. PhD thesis, Brown University, 2001.

L. Peshkin, K.-E. Kim, N. Meuleau, and L. P. Kaelbling. Learning to cooperate via policy search. In *Proceedings of Uncertainty in Artificial Intelligence*, pages 307–314, 2000.

M. Petrik and S. Zilberstein. Average-reward decentralized Markov decision processes. In *Proceedings of the International Joint Conference on Artificial Intelligence*, pages 1997–2002, 2007.

M. Petrik and S. Zilberstein. A bilinear programming approach for multiagent planning. *Journal of Artificial Intelligence Research*, 35(1):235–274, 2009.

J. Pineau and G. Gordon. POMDP planning for robust robot control. In *International Symposium on Robotics Research*, pages 69–82, 2005.

P. Poupart. *Exploiting Structure to Efficiently Solve Large Scale Partially Observable Markov Decision Processes*. PhD thesis, Department of Computer Science, University of Toronto, 2005.

P. Poupart and C. Boutilier. Value-directed compression of POMDPs. In S. T. S. Becker and K. Obermayer, editors, *Advances in Neural Information Processing Systems 15*, pages 1547–1554. MIT Press, 2003.

P. Poupart and C. Boutilier. Bounded finite state controllers. In *Advances in Neural Information Processing Systems 17*, pages 823–830, 2004.

M. L. Puterman. *Markov Decision Processes—Discrete Stochastic Dynamic Programming*. Wiley, 1994.

D. V. Pynadath and S. C. Marsella. Minimal mental models. In *Proceedings of the National Conference on Artificial Intelligence*, pages 1038–1044, 2007.

D. V. Pynadath and M. Tambe. The communicative multiagent team decision problem: Analyzing teamwork theories and models. *Journal of Artificial Intelligence Research*, 16:389–423, 2002.

D. V. Pynadath and M. Tambe. An automated teamwork infrastructure for heterogeneous software agents and humans. *Journal of Autonomous Agents and Multi-Agent Systems*, 7(1-2):71–100, 2003.

Z. Rabinovich, C. V. Goldman, and J. S. Rosenschein. The complexity of multiagent systems: The price of silence. In *Proceedings of the International Conference on Autonomous Agents and Multiagent Systems*, pages 1102–1103, 2003.

R. Radner. Team decision problems. *Annals of Mathematical Statistics*, 33:857–881, 1962.

A. S. Rao and M. P. Georgeff. BDI agents: From theory to practice. In *Proceedings of the First International Conference on Multiagent Systems*, pages 312–319, 1995.

B. Rathnasabapathy, P. Doshi, and P. Gmytrasiewicz. Exact solutions of interactive POMDPs using behavioral equivalence. In *Proceedings of the International Conference on Autonomous Agents and Multiagent Systems*, pages 1025–1032, 2006.

A. Rogers, A. Farinelli, R. Stranders, and N. Jennings. Bounded approximate decentralised coordination via the max-sum algorithm. *Artificial Intelligence*, 175(2):730–759, 2011.

M. Roth, R. Simmons, and M. Veloso. Reasoning about joint beliefs for execution-time communication decisions. In *Proceedings of the International Conference on Autonomous Agents and Multiagent Systems*, pages 786–793, 2005a.

M. Roth, R. Simmons, and M. M. Veloso. Decentralized communication strategies for coordinated multi-agent policies. In L. E. Parker, F. E. Schneider, and A. C. Shultz, editors, *Multi-Robot Systems. From Swarms to Intelligent Automata*, volume III, pages 93–106. Springer, 2005b.

M. Roth, R. Simmons, and M. Veloso. Exploiting factored representations for decentralized execution in multi-agent teams. In *Proceedings of the International Conference on Autonomous Agents and Multiagent Systems*, pages 467–463, 2007.

N. Roy, G. Gordon, and S. Thrun. Planning under uncertainty for reliable health care robotics. In *Proceedings of the International Conference on Field and Service Robotics*, pages 417–426, 2003.

S. Russell and P. Norvig. *Artificial Intelligence: A Modern Approach*. Pearson Education, 3rd edition, 2009.

Y. Satsangi, S. Whiteson, and F. A. Oliehoek. Exploiting submodular value functions for faster dynamic sensor selection. In *Proceedings of the Twenty-Ninth AAAI Conference on Artificial Intelligence*, pages 3356–3363, 2015.

S. Seuken and S. Zilberstein. Memory-bounded dynamic programming for DEC-POMDPs. In *Proceedings of the International Joint Conference on Artificial Intelligence*, pages 2009–2015, 2007a.

S. Seuken and S. Zilberstein. Improved memory-bounded dynamic programming for decentralized POMDPs. In *Proceedings of Uncertainty in Artificial Intelligence*, pages 344–351, 2007b.

S. Seuken and S. Zilberstein. Formal models and algorithms for decentralized decision making under uncertainty. *Journal of Autonomous Agents and Multi-Agent Systems*, 17(2):190–250, 2008.

J. Shen, R. Becker, and V. Lesser. Agent interaction in distributed MDPs and its implications on complexity. In *Proceedings of the International Conference on Autonomous Agents and Multiagent Systems*, pages 529–536, 2006.

E. A. Shieh, A. X. Jiang, A. Yadav, P. Varakantham, and M. Tambe. Unleashing Dec-MDPs in security games: Enabling effective defender teamwork. In *Proceedings of the Twenty-First European Conference on Artificial Intelligence*, pages 819–824, 2014.

Y. Shoham and K. Leyton-Brown. *Multi-Agent Systems*. Cambridge University Press, 2007.

M. P. Singh. *Multiagent Systems: A Theoretical Framework for Intentions, Know-How, and Communications*. Springer, 1994.

S. P. Singh, T. Jaakkola, and M. I. Jordan. Learning without state-estimation in partially observable Markovian decision processes. In *Proceedings of the International Conference on Machine Learning*, pages 284–292, 1994.

T. Smith. *Probabilistic Planning for Robotic Exploration*. PhD thesis, The Robotics Institute, Carnegie Mellon University, 2007.

T. Smith and R. G. Simmons. Heuristic search value iteration for POMDPs. In *Proceedings of the Twentieth Conference on Uncertainty in Artificial Intelligence*, pages 520–527, 2004.

M. T. J. Spaan. Partially observable Markov decision processes. In M. Wiering and M. van Otterlo, editors, *Reinforcement Learning: State of the Art*, pages 387–414. Springer, 2012.

M. T. J. Spaan and F. S. Melo. Interaction-driven Markov games for decentralized multiagent planning under uncertainty. In *Proceedings of the International Conference on Autonomous Agents and Multiagent Systems*, pages 525–532, 2008.

M. T. J. Spaan and N. Vlassis. Perseus: Randomized point-based value iteration for POMDPs. *Journal of Artificial Intelligence Research*, 24:195–220, 2005.

M. T. J. Spaan, G. J. Gordon, and N. Vlassis. Decentralized planning under uncertainty for teams of communicating agents. In *Proceedings of the International Conference on Autonomous Agents and Multiagent Systems*, pages 249–256, 2006.

M. T. J. Spaan, F. A. Oliehoek, and N. Vlassis. Multiagent planning under uncertainty with stochastic communication delays. In *Proceedings of the Eighteenth International Conference on Automated Planning and Scheduling*, pages 338–345, 2008.

M. T. J. Spaan, F. A. Oliehoek, and C. Amato. Scaling up optimal heuristic search in Dec-POMDPs via incremental expansion. In *Proceedings of the Twenty-Second International Joint Conference on Artificial Intelligence*, pages 2027–2032, 2011.

R. St-Aubin, J. Hoey, and C. Boutilier. APRICODD: Approximate policy construction using decision diagrams. In *Advances in Neural Information Processing Systems 13*, pages 1089–1095, 2001.

P. Stone and M. Veloso. Task decomposition, dynamic role assignment, and low-bandwidth communication for real-time strategic teamwork. *Artificial Intelligence*, 110(2):241–273, 1999.

P. Stone and M. Veloso. Multiagent systems: A survey from a machine learning perspective. *Autonomous Robots*, 8(3):345–383, 2000.

R. S. Sutton and A. G. Barto. *Reinforcement Learning: An Introduction*. The MIT Press, 1998.

R. S. Sutton, D. Precup, and S. Singh. Between MDPs and semi-MDPs: A framework for temporal abstraction in reinforcement learning. *Artificial Intelligence*, 112(1):181–211, 1999.

K. P. Sycara. Multiagent systems. *AI Magazine*, 19(2):79–92, 1998.

D. Szer and F. Charpillet. An optimal best-first search algorithm for solving infinite horizon DEC-POMDPs. In *Machine Learning: ECML 2005*, volume 3720 of *Lecture Notes in Computer Science*, pages 389–399, 2005.

D. Szer and F. Charpillet. Point-based dynamic programming for DEC-POMDPs. In *Proceedings of the National Conference on Artificial Intelligence*, pages 1233–1238, 2006.

D. Szer, F. Charpillet, and S. Zilberstein. MAA*: A heuristic search algorithm for solving decentralized POMDPs. In *Proceedings of Uncertainty in Artificial Intelligence*, pages 576–583, 2005.

M. Tambe. Towards flexible teamwork. *Journal of Artificial Intelligence Research*, 7:83–124, 1997.

M. Tambe. *Security and Game Theory: Algorithms, Deployed Systems, Lessons Learned*. Cambridge University Press, 2011.

N. Tao, J. Baxter, and L. Weaver. A multi-agent policy-gradient approach to network routing. In *Proceedings of the International Conference on Machine Learning*, pages 553–560, 2001.

G. Theocharous, K. P. Murphy, and L. P. Kaelbling. Representing hierarchical POMDPs as DBNs for multi-scale robot localization. In *Proceedings of the International Conference on Robotics and Automation*, pages 1045–1051, 2004.

M. Toussaint. Probabilistic inference as a model of planned behavior. *Künstliche Intelligenz*, 3(9):23–29, 2009.

M. Toussaint, S. Harmeling, and A. Storkey. Probabilistic inference for solving (PO)MDPs. Technical report, University of Edinburgh, School of Informatics, 2006.

M. Toussaint, L. Charlin, and P. Poupart. Hierarchical POMDP controller optimization by likelihood maximization. In *Proceedings of the Twenty-Fourth Conference on Uncertainty in Artificial Intelligence*, pages 562–570, 2008.

K. Tuyls and G. Weiss. Multiagent learning: Basics, challenges, and prospects. *AI Magazine*, 33(3):41–52, 2012.

K. Tuyls, P. J. Hoen, and B. Vanschoenwinkel. An evolutionary dynamical analysis of multi-agent learning in iterated games. *Journal of Autonomous Agents and Multi-Agent Systems*, 12(1):115–153, 2006.

P. Varaiya and J. Walrand. On delayed sharing patterns. *IEEE Transactions on Automatic Control*, 23(3):443–445, 1978.

P. Varakantham, R. Nair, M. Tambe, and M. Yokoo. Winning back the cup for distributed POMDPs: Planning over continuous belief spaces. In *Proceedings of the International Conference on Autonomous Agents and Multiagent Systems*, pages 289–296, 2006.

P. Varakantham, R. Maheswaran, T. Gupta, and M. Tambe. Towards efficient computation of quality bounded solutions in POMDPs. In *Proceedings of the Twentieth International Joint Conference on Artificial Intelligence*, pages 2638–2643, 2007a.

P. Varakantham, J. Marecki, Y. Yabu, M. Tambe, and M. Yokoo. Letting loose a SPIDER on a network of POMDPs: Generating quality guaranteed policies. In *Proceedings of the International Conference on Autonomous Agents and Multiagent Systems*, pages 817–824, 2007b.

P. Varakantham, J.-y. Kwak, M. E. Taylor, J. Marecki, P. Scerri, and M. Tambe. Exploiting coordination locales in distributed POMDPs via social model shaping. In *Proceedings of the International Conference on Automated Planning and Scheduling*, pages 313–320, 2009.

P. Varshavskaya, L. P. Kaelbling, and D. Rus. Automated design of adaptive controllers for modular robots using reinforcement learning. *International Journal of Robotics Research*, 27(3-4):505–526, 2008.

P. Velagapudi, P. Varakantham, P. Scerri, and K. Sycara. Distributed model shaping for scaling to decentralized POMDPs with hundreds of agents. In *Proceedings of the International Conference on Autonomous Agents and Multiagent Systems*, pages 955–962, 2011.

N. Vlassis. *A Concise Introduction to Multiagent Systems and Distributed Artificial Intelligence*. Synthesis Lectures on Artificial Intelligence and Machine Learning. Morgan & Claypool, 2007.

N. Vlassis, M. L. Littman, and D. Barber. On the computational complexity of stochastic controller optimization in POMDPs. *ACM Transactions on Computation Theory*, 4(4):12:1–12:8, 2012.

J. von Neumann and O. Morgenstern. *The Theory of Games and Economic Behavior*. Princeton University Press, 1944.

M. de Weerdt and B. Clement. Introduction to planning in multiagent systems. *Multiagent and Grid Systems*, 5(4):345–355, 2009.

M. de Weerdt, A. ter Mors, and C. Witteveen. Multi-agent planning: An introduction to planning and coordination. In *Handouts of the European Agent Summer School*, pages 1–32, 2005.

G. Weiss, editor. *Multiagent Systems: A Modern Approach to Distributed Artificial Intelligence*. MIT Press, 1999.

G. Weiss, editor. *Multiagent Systems*. MIT Press, 2nd edition, 2013.

M. Wiering. Multi-agent reinforcement learning for traffic light control. In *Proceedings of the International Conference on Machine Learning*, pages 1151–1158, 2000.

M. Wiering and M. van Otterlo, editors. *Reinforcement Learning: State of the Art*. Adaptation, Learning, and Optimization. Springer, 2012.

M. Wiering, J. Vreeken, J. van Veenen, and A. Koopman. Simulation and optimization of traffic in a city. In *IEEE Intelligent Vehicles Symposium*, pages 453–458, 2004.

A. J. Wiggers, F. A. Oliehoek, and D. M. Roijers. Structure in the value function of zero-sum games of incomplete information. In *Proceedings of the AAMAS Workshop on Multi-Agent Sequential Decision Making in Uncertain Domains (MSDM)*, 2015.

K. Winstein and H. Balakrishnan. TCP ex machina: Computer-generated congestion control. In *SIGCOMM*, pages 123–134, 2013.

H. S. Witsenhausen. Separation of estimation and control for discrete time systems. *Proceedings of the IEEE*, 59(11):1557–1566, 1971.

S. Witwicki and E. Durfee. From policies to influences: A framework for nonlocal abstraction in transition-dependent Dec-POMDP agents. In *Proceedings of the International Conference on Autonomous Agents and Multiagent Systems*, pages 1397–1398, 2010a.

S. Witwicki, F. A. Oliehoek, and L. P. Kaelbling. Heuristic search of multiagent influence space. In *Proceedings of the Eleventh International Conference on Autonomous Agents and Multiagent Systems*, pages 973–981, 2012.

S. J. Witwicki. *Abstracting Influences for Efficient Multiagent Coordination Under Uncertainty*. PhD thesis, University of Michigan, 2011.

S. J. Witwicki and E. H. Durfee. Flexible approximation of structured interactions in decentralized Markov decision processes. In *Proceedings of the International Conference on Autonomous Agents and Multiagent Systems*, pages 1251–1252, 2009.

S. J. Witwicki and E. H. Durfee. Influence-based policy abstraction for weakly-coupled Dec-POMDPs. In *Proceedings of the International Conference on Automated Planning and Scheduling*, pages 185–192, 2010b.

S. J. Witwicki and E. H. Durfee. Towards a unifying characterization for quantifying weak coupling in Dec-POMDPs. In *Proceedings of the Tenth International Conference on Autonomous Agents and Multiagent Systems*, pages 29–36, 2011.

M. Wooldridge. *An Introduction to MultiAgent Systems*. Wiley, 2002.

M. Wooldridge and N. R. Jennings. Intelligent agents: Theory and practice. *Knowledge Engineering Review*, 10(2):115–152, 1995.

F. Wu, S. Zilberstein, and X. Chen. Point-based policy generation for decentralized POMDPs. In *Proceedings of the International Conference on Autonomous Agents and Multiagent Systems*, pages 1307–1314, 2010a.

F. Wu, S. Zilberstein, and X. Chen. Rollout sampling policy iteration for decentralized POMDPs. In *Proceedings of Uncertainty in Artificial Intelligence*, pages 666–673, 2010b.

F. Wu, S. Zilberstein, and N. R. Jennings. Monte-Carlo expectation maximization for decentralized POMDPs. In *Proceedings of the Twenty-Third International Joint Conference on Artificial Intelligence*, pages 397–403, 2013.

M. Wunder, M. L. Littman, and M. Babes. Classes of multiagent Q-learning dynamics with epsilon-greedy exploration. In *Proceedings of the International Conference on Machine Learning*, pages 1167–1174, 2010.

P. Xuan, V. Lesser, and S. Zilberstein. Communication decisions in multi-agent cooperation: Model and experiments. In *Proceedings of the International Conference on Autonomous Agents*, pages 616–623, 2001.

M. Yokoo. *Distributed Constraint Satisfaction: Foundation of Cooperation in Multiagent Systems*. Springer, 2001.

M. Yokoo and K. Hirayama. Distributed breakout algorithm for solving distributed constraint satisfaction problems. In *Proceedings of the Second International Conference on Multiagent Systems*, pages 401–408, 1996.

W. Zajdel, A. Taylancemgil, and B. A. Kröse. Dynamic Bayesian networks for visual surveillance with distributed cameras. In *EuroSSC*, pages 240–243, 2006.

Y. Zeng, P. Doshi, Y. Pan, H. Mao, M. Chandrasekaran, and J. Luo. Utilizing partial policies for identifying equivalence of behavioral models. In *Proceedings of the National Conference on Artificial Intelligence*, pages 1083–1088, 2011.

E. Zermelo. Uber eine Anwendung der Mengenlehre auf die Theorie des Schachspiels. In *Proceedings of the Fifth International Congress of Mathematicians II*, pages 501–504, 1913.

L. S. Zettlemoyer, B. Milch, and L. P. Kaelbling. Multi-agent filtering with infinitely nested beliefs. In *Advances in Neural Information Processing Systems 21*, pages 1905–1912, 2009.

C. Zhang, V. Lesser, and S. Abdallah. Self-organization for coordinating decentralized reinforcement learning. In *Proceedings of the International Conference on Autonomous Agents and Multiagent Systems*, pages 739–746, 2010.

Printed in the United States
By Bookmasters